The Canyon Revisited

The Canyon Revisited

A REPHOTOGRAPHY OF THE Grand Canyon 1923/1991

Donald L. Baars
Rex C. Buchanan

REPHOTOGRAPHY BY
John R. Charlton

University of Utah Press
Salt Lake City

∞ Printed on acid-free paper

Photographs from the Emery Kolb collection reproduced by permission of the Cline Library, Northern Arizona University.

Photographs and drawings from the R. C. Moore Collection, University of Kansas Archives, are reproduced courtesy of the Spencer Research Library, University of Kansas.

See page 168 for the sources of the historical photographs.

LIBRARY OF CONGRESS CATALOGING-IN-PUBLICATION DATA

Baars, Donald L.
The canyon revisited : a rephotography of the Grand Canyon, 1923/1991 / Donald L. Baars, Rex C. Buchanan ; rephotography by John R. Charlton.
p. cm.
Includes bibliographical references (p.) and index.
ISBN 0-87480-458-2 (acid-free paper)
1. Grand Canyon (Ariz.)—Geography. 2. Grand Canyon (Ariz.)—Pictorial works. 3. Geology—Arizona—Grand Canyon. 4. Photography—Arizona—Grand Canyon. I. Buchanan, Rex, 1953– . II. Charlton, John R. III. Title.
F788.B2 1994
917.91′32—dc20 94-25685

CONTENTS

Photographing Grand Canyon

On the morning of August 1, 1923, eleven men in five boats pushed off from Lees Ferry, Arizona, into the Colorado River. Led by Claude H. Birdseye, chief topographic engineer for the United States Geological Survey, they made up one of the most important scientific surveys ever undertaken along the Colorado. Their monumental task: to map Marble and Grand canyons as they surveyed for possible dam and reservoir sites—while avoiding disaster on a violent river that still challenges travelers in modern rafts.

In 1989, we discovered a large, tattered album of photographs from Birdseye's 1923 expedition left in the University of Kansas Archives by Raymond C. Moore, the expedition's geologist. Entranced, we wondered what Moore would see and think if he were to repeat his journey today.

We also wondered what rephotographing the older images would tell us about how natural and human forces had altered the Grand Canyon between 1923 and the present—a time in which dams had been erected across the Colorado and human traffic down the river had increased dramatically. By going back to where the original pictures were taken, photographing the landscape as it looks today, and then comparing the new images with the old, we could document ways in which the canyon had changed over the past sixty-eight years.

Our curiosity gave birth to a trip that in 1991 followed in Birdseye's wake and rephotographed some of the scenes his party had captured. This book shows the results, displaying the original photographs alongside the new so that changes—or the lack of them—are apparent. Among other things, the pairs of photographs show the overwhelming impact humans have had on the Grand Canyon.

Before 1923, only ten expeditions of note had ventured into the canyon. The first two were John Wesley Powell's scientific adventures of 1869 and 1872; three members of the 1869 crew climbed out at Separation Canyon, only to be killed by Indians, and the 1872 party left the canyon via Kanab Creek. Next came a railroad survey led by Robert Brewster Stanton in 1889–90, during which three men drowned. Two parties, those of Nathaniel Galloway and William Richmond, made the trip in 1896, following closely on the heels of George Flavell and Ramón Montez. John King, Arthur Sanger, and Elias B. "Hum" Wooley came next, in 1903, and Galloway made a second trip, along with Julius Stone, in 1909. Charles Russell and Edwin R. Monett ran the river in 1907–1908, followed by the photographic expedition of brothers Ellsworth and Emery Kolb in 1911. Russell made a nearly disastrous second trip in 1914; his party climbed out of the canyon defeated and broken at Bass Trail.

In all, only forty-five men had made successful trips past Bright Angel Creek, eighty-eight miles downstream from Lees Ferry—and of those, only twenty-nine had completed the rigorous journey all the way through the canyon to the mouth of the Virgin River. Preceded by so little river traffic, Birdseye's survey crew traveled a canyon untouched except by American Indians who had visited there for centuries.

Even as late as 1949, only a hundred people had floated the canyon. River camps for tourists built by Seth Tanner, John Hance, William Bass, and "The Hermit" Boucher (pronounced Boo-shay) had sprung up in the 1890s; a hotel opened near the mouth of Diamond Creek in 1894 but closed five years later. These amenities brought tourists to the canyon floor, but only in isolated spots.

Today, as many as seventy-five people depart just from Lees Ferry every day between April and September. It is not uncommon to see forty-five people or more on a single forty-five-foot J-rig or triple-pontoon rig. In 1969, the Four Corners Geological Society celebrated the Powell centennial with a trip that took 156 geologists down the Colorado in seventeen boats—the largest single trip ever conducted in the canyon. Such events prompted the National Park Service to impose strict regulations on the size and number of trips permitted, as well as sanitation and safety standards.

By 1991 some twenty thousand people a year were traversing the canyon by boat. Thousands more tourists rode mules to Phantom Ranch near the mouth of Bright Angel Creek, and nude sunbathers decorated the foot of every hiking trail from the rim to the river. Although strict regulations and the travelers' desire to cooperate kept the canyons clean, hundreds of thousands of visitors had trampled every inch of sand, stepped on fragile vegetation, and rolled boulders to and fro. The canyon floor seemed less a pristine wilderness than a typical national park.

WHILE HUMAN TRAFFIC in the canyon was burgeoning, the enormous impact of dams also hit the Colorado hard. When Birdseye's expedition set out in 1923, there were no dams on the Colorado River. After running nearly dry at some times of the year, the river could at other times be scoured by raging torrents of up to 300,000 cubic feet per second (cfs).

River water was never clean. It ran red most of the time (*colorado* means "red" or "colorful" in Spanish) as eroded sediments funneled into it from the entire Colorado Plateau and from much of the Rocky Mountains of Colorado and Wyoming. The flow could consist of as much as 80 percent silt when discharge rates were high, and the median annual sediment concentration rate was 1,500 parts per million (Turner and Karpiscak 1980). An old Park Service cliché called the river "too thick to drink and too thin to plow." Its sediments, mostly sand, mud, and silt, were sluiced into the head of the Gulf of California by the millions of tons, where they gradually filled in the head of the gulf and formed rich farmlands in southeastern California.

In 1935, the construction of Hoover Dam and its associated reservoir, Lake Mead, set off a chain reaction of engineering feats that changed the Colorado in ways no one anticipated. Hoover Dam trapped the river's voluminous sedimentary load in Lake Mead, and soon that artificial basin began to fill. Something had to be done to save the dam. The obvious solution was to build more dams upstream. Flaming Gorge Dam (locals call it "Flaming George") went in across the upper Green River in 1962, along with Navajo Dam across the San Juan River. They were followed by Glen Canyon Dam in 1963 and the Curecanti Unit dams in Colorado in 1965 and 1966.

Of all these dams, by far the most significant for Marble and Grand canyons was Glen Canyon Dam above Lees Ferry. Its reservoir, Lake Powell, now receives sediments once destined for Lake Mead, so that water entering Marble Canyon is clear. New sediment is added only by minor influxes from the Paria and Little Colorado rivers and by flash floods in smaller tributaries. The dam holds annual river flow relatively constant; flood waters have been all but eliminated. The river's flow is dictated by the electrical needs of cities in southern Arizona and California and rarely exceeds 35,000 cfs.

Still, daily fluctuations of 5,000 to 25,000 cfs reflect hourly power needs at the dam and vex boaters with what is known as "the Udall effect" (so named because the dam was built when Stewart Udall was secretary of the interior). Boats not properly positioned during the evening's high water or moved during the night may be draped high and dry across rocky ridges the next morning.

The overall result of damming the river at Glen Canyon is that sediments once deposited in Marble and Grand canyons are now being eroded from the upper canyons and redeposited locally in the lower canyons below upper Granite Gorge or flushed into Lake Mead. In human terms, the Colorado River "wants" to carry a certain amount of sediment. Because Glen Canyon Dam removes much of it, the river picks up silt below the dam, often by eroding beaches. Shorelines are no longer scoured by floodwater, and so the riparian environment has been lowered topographically. New vegetation flourishes in places that previously were flooded. Introduced species such as tamarisk (salt cedar) now dominate the shoreline. Along with the new vegetation comes an entire change in the ecology near the river. Insects, birds, small mammals—all are affected by the changing shoreline. Even once-exotic fish such as trout thrive in the cleaner, colder water that is discharged from Lake Powell.

THE BIRDSEYE EXPEDITION of 1923 was the vanguard of the army of engineers who would dam the Colorado. The already lucrative agricultural lowlands downstream would become even more profitable if annual flooding could be controlled. Although Stanton had surveyed a possible railroad route in 1889, the canyons had never been thoroughly mapped, and the USGS had no topographic or geological survey upon which to base damsite decisions. Birdseye's was the first truly successful survey of the Colorado River in Marble and Grand canyons.

Comparing the Birdseye expedition to John Wesley Powell's trip of 1869, historian David Lavender wrote:

> Scientifically, this was the most significant trip since Powell's. Thanks to Roger Birdseye's efficient supply system, they were able to stay in

> the Grand Canyon for a longer period than earlier parties had. As a result, Raymond Moore had found time to broaden the geologic knowledge of the canyon. The surveyors would produce from their figures the first topographic map of much of the canyon bottom. In addition, they had examined twenty-one potential dam sites, an essential step toward what they called "the conquest of the Colorado." (1985:65)

Because the most essential ingredient of dam construction is the integrity of the dam's rock abutments, the survey needed a capable geologist. Raymond C. Moore, a preeminent geologist and invertebrate paleontologist, was appointed to fill this important assignment. In Kansas, Moore had built one of the country's outstanding state geological surveys, which he headed from 1916 until he stepped down in 1954.

Three members of the expedition took the pictures we later attempted to repeat. The party's official photographer, Eugene C. LaRue, was also a hydraulic engineer who served as chief hydrologist. His photos were the most important ones to the survey itself and were used to illustrate the reports he and R. C. Moore published in 1925.

A second photographer was Emery Kolb, the head boatman. Perhaps the most important person to the safety and well-being of the party, Kolb was the only member of the group to have previously run the length of the canyons. Together with his brother Ellsworth, Kolb became the first commercial photographer of the Grand Canyon, and by 1923 he was a nationally known photographer and explorer. His strongly individual character and his commercial interest in photographing the canyon often clashed with the serious agenda of the government surveyors. He repeatedly argued with Claude Birdseye about conflicts between his boatman duties and his interests in photographing the Grand Canyon. But Kolb managed both to get the expedition downriver successfully and to make his own photographs using a hand-held camera.

Lewis R. Freeman, a journalist who signed on as a boatman to do an article for *National Geographic* magazine, was the third photographer on the 1923 expedition. His account in the May 1924 *National Geographic* and in later books highlighted the interactions of the people on the trip. His pictures, usually taken at campsites, offer a photojournalistic record of a historic expedition that encompassed several strong personalities.

Claude Birdseye and his party left the Colorado River at Needles, California, in October 1923. In their three months on the river, Birdseye's crew had measured elevations and drawn geologic cross sections throughout the length of the Grand Canyon and Marble Canyon. But without realizing it, his men had accomplished something else that was equally important: they recorded the canyon before it was swamped by tourists or threatened by pollution, and before the Colorado was dammed. In short, they provided a benchmark record of the world's grandest natural spectacle in its pristine state.

LEE C. GERHARD, state geologist and director of today's Kansas Geological Survey (KGS), recognized the importance of a rephotographic survey that would compare the present-day canyon to Birdseye's benchmark. Gerhard recruited Kansas state senator Frank Gaines to provide personal financial backing for the project. Don Baars, senior geologist at KGS, was appointed the trip's technical leader, for his experience as a geologist and river guide in Grand Canyon would be useful in managing the logistics of preparation. Rex Buchanan, a historian and the assistant director for publications and public affairs at KGS, was to provide historical background and keep running records of our day-to-day achievements and trials. The task of carrying out the repeat photography fell to KGS staff photographer John Charlton.

The real heroes of the project were the staff and crew of Ted Hatch River Expeditions. As the exigencies of photography made our daily progress and campsites unpredictable, we demanded of the crew patience and endurance far beyond that needed for ordinary tourist trips. Bill Ellwanger, the lead boatman, with more than two hundred previous trips through the canyons, was invaluable in helping us locate the most difficult of the 1923 camera stations. Bill's wife, Patti, and J. P. Running provided fine river cuisine and loving hospitality. And thanks to the planning and ingenuity of Ted Hatch, we never ran low on fresh food and much-coveted ice.

As he planned the trip, Don Baars continued searching for more of the original photographs from the 1923 survey. In the Special Collections Library at Northern Arizona University, he found the Emery Kolb collection that housed more than five hundred pictures—two hundred of which might possibly be rephotographed. Of those, we selected eighty of the best images to try to reshoot.

Among other sources for the old photos were archives that allowed varied access to the original negatives or prints. The best quality prints came from R. C. Moore's collection, courtesy of the University of Kansas Archives. The photographs from the earlier trip varied in archival quality as well as in photographic technique and point of view. Often we found no record which member of the 1923 crew had taken a particular photograph, and we could distinguish them only by what we knew of the photographers' different interests in subject matter and preferences for vantage points.

We chose photographs from the 1923 archives that represented important aspects of Birdseye's expedition in a series we thought we could achieve during our time on the river. We anticipated that we would not be able to find some of the pictured locations and that others would not have changed in any obvious way since Birdseye passed through. But we felt confident we would rediscover enough of the original vantage points to document what had happened to the canyons over sixty-eight years.

We departed Lees Ferry on September 8, 1991, approximately midpoint in that segment of the annual sunlight cycle experienced by the 1923 crew. Our schedule called for us to rephotograph in some two weeks an area representing three months' worth of travel and study in 1923. In sixteen arduous days we replicated forty-five of the hundreds of historic photographs of 1923.

Ours was not the first rephotographic survey of the Colorado River in the Grand Canyon. The canyon makes a good subject because its dramatic geology is one of the most often-photographed features on the continent. Scientists, historians, and photographers often use repeat photography to document and evaluate landscape changes there as elsewhere. The premises and techniques for using repeat photography in geological studies are elaborated in Harold Malde's 1973 article, "Geologic Bench Marks by Terrestrial Photography." The best-known examples of repeat photography are those in *Second View: The Rephotographic Survey Project,* by Mark Klett and others (1984), in which famous nineteenth-century photographs of the West are reproduced alongside contemporary repeat images.

Two previously published rephotographic studies of the Colorado River in the Grand Canyon are the technical report *Recent Vegetation Changes along the Colorado River between Glen Canyon Dam and Lake Mead, Arizona,* by Raymond Turner and Martin Karpiscak (1980), and the more popular *In the Footsteps of John Wesley Powell,* by Hal Stephens and Eugene Shoemaker (1987). The first study contains a broad range of earlier photographs, from those taken during Powell's second expedition to government photographs taken in the 1960s after the completion of Glen Canyon Dam. Its purpose was to study changes in vegetation resulting from river regulation.

Our project has more in common with *In the Footsteps of John Wesley Powell.* Much as we did, Stephens and Shoemaker followed a famous scientific expedition—Powell's two-year second journey—and spent three months in 1968 rephotographing Powell's locations. Another, more recent rephotographic survey, *Historical Channel Change of Kanab Creek,* by Robert H. Webb, Spence S. Smith, and V. Alexander S. McCord (1992), repeats the photos taken by John Hillers, a member of Powell's party, on Kanab Creek, a major tributary to the Colorado River.

For the rephotography in our 1991 project, we used Polaroid positive/negative film because it provided proofs that we could compare on the spot with our copies of the 1923 photographs. When the earlier photograph and the new proof lined up as closely as possible, we exposed the negatives, identifying and storing them for later development at the Kansas Geological Survey. Previous experience with such negatives in extreme climatic conditions had led us to believe this procedure was prudent, and in fact we suffered only a few casualties owing to development problems with the film.

As a backup color medium, we used medium-format roll film—a method that worked well until the viewing glass for the roll-back attachment on the view camera broke in a backpack on the fifth day of the trip. Bill Ellwanger's resourcefulness proved invaluable. He noticed that the remnant of broken glass was about the size of some plastic audiocassette cases he had on the raft. The plastic fit perfectly on both sides of the delicate shard, protecting the glass and permitting us to view on it through the plastic.

Because we initially determined each vantage point in the view glass of our standard four-by-five-inch camera, along with the instant-film back, we could then substitute the appropriate lens and roll-film back after the camera and tripod were already positioned for the repeat shot. The image could then be focused on the small remaining piece of the broken view glass without being framed again. Some additional adjustments of aperture and exposure gave us the roll-film backup images we needed despite having broken the ground glass.

PHOTOGRAPHIC TECHNOLOGY ASIDE, it was rediscovering the vantage points of the 1923 photos that challenged us most. Finding the true vantage point—the original photographer's exact location and camera angle—allows direct comparison between the original and the repeat photograph. It is also important to duplicate as precisely as possible the time of year and day when the original photograph was taken and to get the same illumination from sunlight on the subject. When all these factors are accurately reproduced, the effect of the repeat image is visually remarkable. Viewers can concentrate on the differences or similarities between the old and new images without being misled by discrepancies in shadows or angles or points of view.

If the original vantage point or the subject itself has changed, however, the rephotographer may have to make a calculated readjustment of the vantage point in order to take a meaningful photo. In our case, the unpredictability of travel on the Colorado River sometimes made it impossible to repeat a shot on the same day or at the same hour the original was taken. We do have virtually identical lighting in some pairs because we tried to schedule our project as closely as possible to the same time of year as the 1923 survey. Moreover, rephotographic river expeditions in the Grand Canyon often have to settle for less than ideal sunlight conditions. All these variables must be identified and factored into the comparison of the old and new images. Our captions include not only times of day for our repeat photographs but also estimated times for the 1923 images, based upon their similarities to and differences from ours.

Discovering vantage points turned out to be complicated—indeed, our searches consumed the greatest portion of our time. Repeat photographs reveal what has changed in front of the camera but only suggest what has changed on the actual spot where each photograph was made. We sometimes found that the vantage points had changed more dramatically than the scenes displayed in the photographic pairs.

The horizon in a landscape is normally the easiest element to reestablish with perfect accuracy, but a mile down inside the Grand Canyon it is atop a vertical column. In effect, the horizon begins at river level (seen upside down in a view-camera glass) and reaches up thousands of feet to the top of the canyon walls. The complexity of this background is unforgiving, but it does offer a great deal of visual information for locating the vantage point. If we first found the horizontal location of the original photo in relation to the river, we could then establish the vertical position of the image, often by climbing the canyon walls.

Despite the constant canyon features, we found the river foreground confusing. In some places the shoreline along the river had changed so much as to be virtually unrecognizable from the older photograph. It was hard to be sure whether we stood on the right, if altered, spot, or whether we were at the wrong vantage point. Sometimes we had to conclude that the original vantage point was now behind a forest of tamarisk or buried beneath a landslide, or that it had sat atop a boulder or on a rock ledge that no longer existed. When this happened, we either made as close an approximation to the original vantage point as we could, or we simply abandoned the view and went on to the next site.

Every day the canyon dished out a new set of vantage-point challenges. Each vantage point on the survey had its own story, although not all those stories are necessary to understand the photographs. The vantage locations are important for future surveys: once a rephotographer discovers a vantage point and notes it on a map, the location may be found more easily by later rephotographic projects. Or next time it may not exist at all.

THE PHOTOGRAPHS MADE on the 1923 USGS expedition are unique: they were the last government-survey photographs taken prior to the construction of Hoover Dam in the 1930s and Glen Canyon Dam in the 1960s. Only one of the 1923 images rephotographed in our series has ever before been published in a repeat set—and even it demonstrates that different repeat photographs of a scene will produce different results. The real scientific and historical value of repeat photography lies not simply in comparing two images but rather in comparing a series of photographs made over time from the same vantage point.

Each of the three photographers on the 1923 survey used a different kind of camera and photographed for a different purpose. Eugene LaRue did not like heights and usually photographed potential dam sites along the river rapids and sidestreams. He often used a panoramic camera mounted on a tripod. We did not reproduce his panoramic images but repeated other photographs that he took with a more normal-format camera. These still required a wide-angle lens on a contemporary view camera. The formats of both Freeman's and Kolb's cameras turned out to be repeatable with a normal lens.

Emery Kolb liked to scale canyon walls and climb out on slick rocks and narrow ledges—the vantage points of a longtime river runner searching for a scene never before photographed. His images are usually views up and down the river taken from high locations, and they were often the most difficult to find and dangerous to rephotograph. Luckily, Kolb had a kindred spirit on our trip. Bill Ellwanger used his lifelong knowledge of the canyon to help us find vantage points; sometimes he was the sole photographic assistant at sites where no one else would follow.

Lewis Freeman's journalistic photos came to our aid by incorporating human figures that provided a scale. We too tried to incorporate a human element into our rephotographic images. Freeman's photographs portray a group of professional scientists at the beginning of the trip who gradually came to look more like a grizzled band of castaways. Our group started out in modern, motorized, neoprene rafts, wearing Patagonia clothing and drinking iced cocktails in the evening. But in the group photo taken at the end of the trip, we do not look like the same people pictured in the first photo taken at Lees Ferry. Indeed, we were not.

Today, people normally think of photographs made in the Grand Canyon as scenic and pictorial. That kind of image was not the intent of the 1923 survey photographers, nor was it ours. LaRue's images were meant only to document potential dam sites, although it is apparent that his interest in photography became keener as he journeyed deeper into the canyon. The strategic and rather technical pictures the photographers took along Badger and Soap Creek rapids and at other locations in Marble Canyon (photo pairs 2–8) seem to reflect the group's earnest mood at the beginning of their survey. In later photographs, a freer spirit emerges (photo pair 21) as the crew members learned how to survive the river's perils. The timelessness of the canyon must surely have affected them even as it did us.

Along with still photographs, LaRue also used sixteen-millimeter motion-picture film to document the 1923 survey. The government later incorporated this footage into a film promoting dam construction for irrigation in California's Imperial Valley. By that time LaRue had left the USGS over repeated disputes with the government's policies for locating dam sites on the Colorado River:

> Eugene LaRue lost his job with the USGS because he could not hold his tongue. Even after the Bureau of Reclamation had decided to control the Colorado River by means of a high dam in either Boulder or Black canyon, he

> kept insisting, publicly and stridently, that a better site was the one he had located in Glen Canyon, four miles above Lee's Ferry. His superiors told him to shut up or get out. Stubbornly turning his back on fifteen years of hydrological work along the river, he left government service and became a private consultant. (Lavender 1985:66)

Comparing the photographs from the Powell expedition with those from the 1923 survey, we see much the same kinds of environmental conditions along the river. There is an awesome, wild quality common to both sets that makes any human presence seem precarious and endangered. We see flash floods changing the configurations of rapids. We see the river scouring the canyon walls of vegetation and depositing new expanses of beach sand.

Comparable images from 1991 show a river along which unreplenished beaches are filling with increasing vegetation. The rapids now change only when Glen Canyon Dam overflows. Tourists, scientists, and park rangers travel the dammed and domesticated river.

As impressed as we were by the evidence of human impact along the Colorado, we recognized the almost complete lack of change in gross aspects of canyon geomorphology. To be sure, we found small geological changes such as rock falls and canyon erosion having nothing to do with river-flow regulation—for example the rotting ledges of the vantage points at Nankoweap Ruins [photo pair 17] and Vulcan's Anvil [photo pair 41]. But sixty-eight years is a mere moment in the geological history of the canyons, which have had the advantage of a few million years of advance preparation. The reconfiguration of a couple of rapids by tributary mudflows pales in the overall scheme of things.

In the brief time since Birdseye, Moore, and company traversed the Colorado, human impact has been dramatic. It is our hope that the photographic pairs presented in this book will show clearly the changes in the canyon, and that they honor the memories of the men who made the trip in 1923.

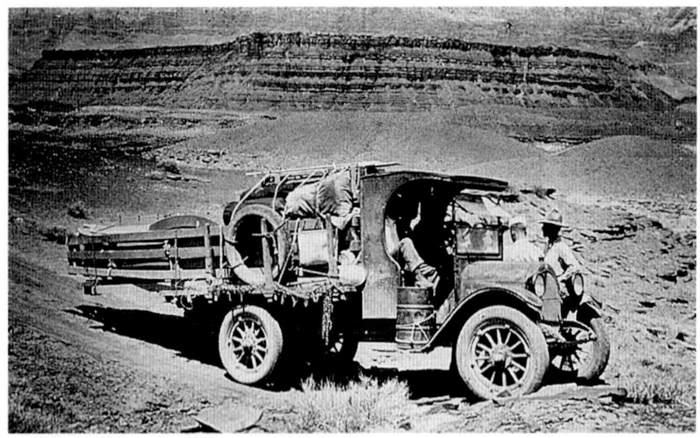

A truck hauls one of the boats to Lees Ferry (R. C. Moore Collection).

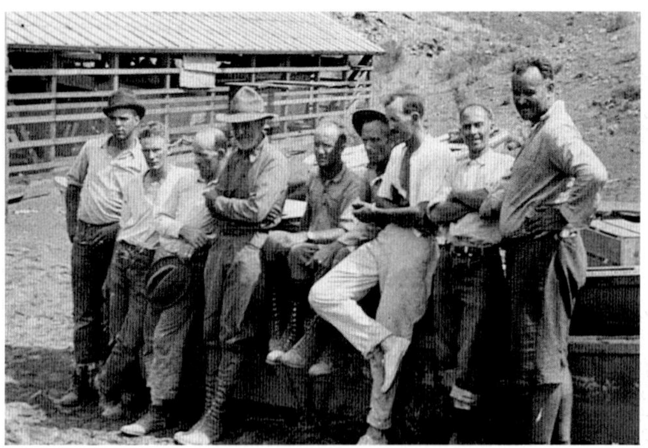

The 1923 crew. From left: Leigh Lint, H. E. Blake, Frank Word, C. H. Birdseye, R. C. Moore, R. W. Burchard, E. C. LaRue, L. R. Freeman, and Emery Kolb (R. C. Moore Collection).

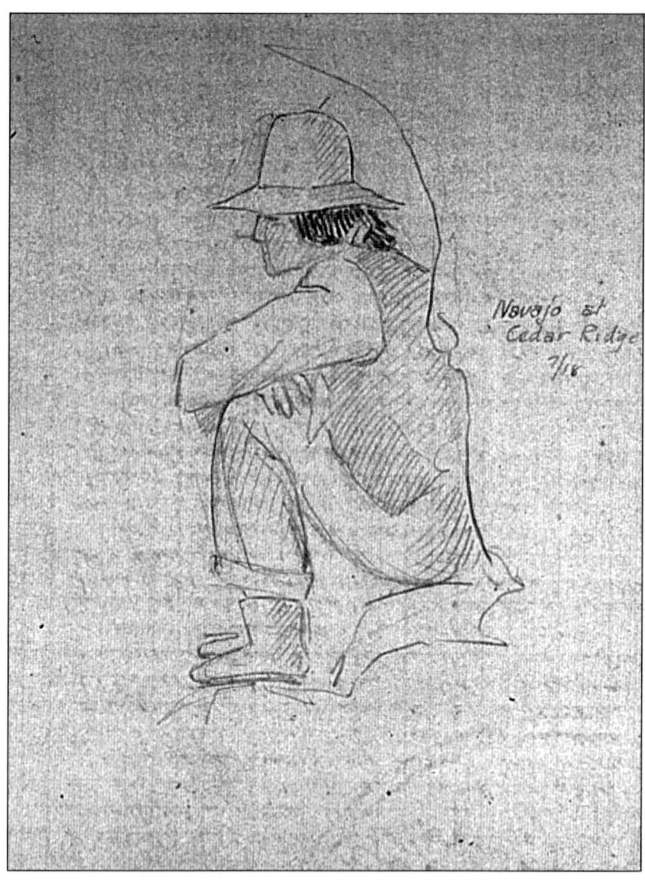

R. C. Moore's sketch of a Navajo near Cedar Ridge, on the way to Lees Ferry (from Moore's Grand Canyon notebook, R. C. Moore Collection).

Two Trips down the Canyon

Rex Buchanan

God dont lie.
No, said the judge. He does not. And these are his words.
He held up a chunk of rock.
He speaks in stones and trees, the bones of things.

—Cormac McCarthy
Blood Meridian

SEPTEMBER 8, 1991: This morning we leave from Lees Ferry, the same departure point as that of the 1923 crew. The parking lot here is paved over now and far more heavily trafficked on this Sunday morning. The Colorado River is clear and cold, the green of a Heineken beer bottle; much of its silt settles out at Lake Powell, a few miles above us. When the 1923 crew left here on that hot August day, Lake Powell didn't exist, and the Colorado River was characteristically muddy.

The 1923 crew was composed of eleven members, including four boatmen. Emery Kolb, the head boatman, was the only one of the group who had previously been through the canyon. The expedition—financed by the U.S. Geological Survey (USGS) and two private power companies, Southern California Edison and Utah Power and Light—was to continue work that had started two years before with surveying expeditions on the San Juan and Green rivers, tributaries of the Colorado. But the 1923 trip was the big one, the one that would tackle the grandest canyon of them all. To make it, the crew took four wooden boats (the *Marble,* the *Glen,* the *Grand,* and the *Boulder*) and one canvas boat (the *Mojave*).

Our group has sixteen members—most of us connected with the Kansas Geological Survey—and is carried on two thirty-three-foot motorized rubber rafts, huge pontoons on either side, operated by Hatch River Expeditions, a commercial river-running outfit.

Less than a mile below Lees Ferry, the Paria River meets the Colorado and turns the water chocolate brown. It rained last night and the side streams are dirty with silt. At mile 8, we hit our first rapid of the trip: Badger Creek Rapid. On the river runner's scale of one to ten, with ten being the biggies, Badger Creek Rapid rates a measly three. Birdseye's group camped here on their first night out; late evening rain cascaded over the canyon walls. On the first day of our trip, skies are cloudless and our ride through Badger Creek is quick. People near the front of the boat get wet, those at the back stay dry. We pull over to the west side of the river, then hike back along the bank to shoot our first repeat photograph of the trip (photo pair 2), taking about two hours. On the way to our next stop, we spot the trip's first bighorn sheep—a ram whose brownish gray hide blends with the red and black desert-varnished rocks in the canyon wall, his horns nearly a full curl.

Back upstream, Glen Canyon Dam removes silt from the river for a time, so that the Colorado now carves sediment away from the sandy beaches along the canyon. In the thirty years since the dam's construction, many beaches have been eaten away, changing the canyon's geology and ecology. That becomes clear when we move on for a photo of 10-Mile Rock, a huge limestone boulder in the river's midst (photo pair 3). We can't match exactly the 1923 vantage point because much of the sand beach is gone.

We make camp at Soap Creek. We are likely very close to the place where the 1923 group camped and set up their radio—after considerable discussion about whether the receiver

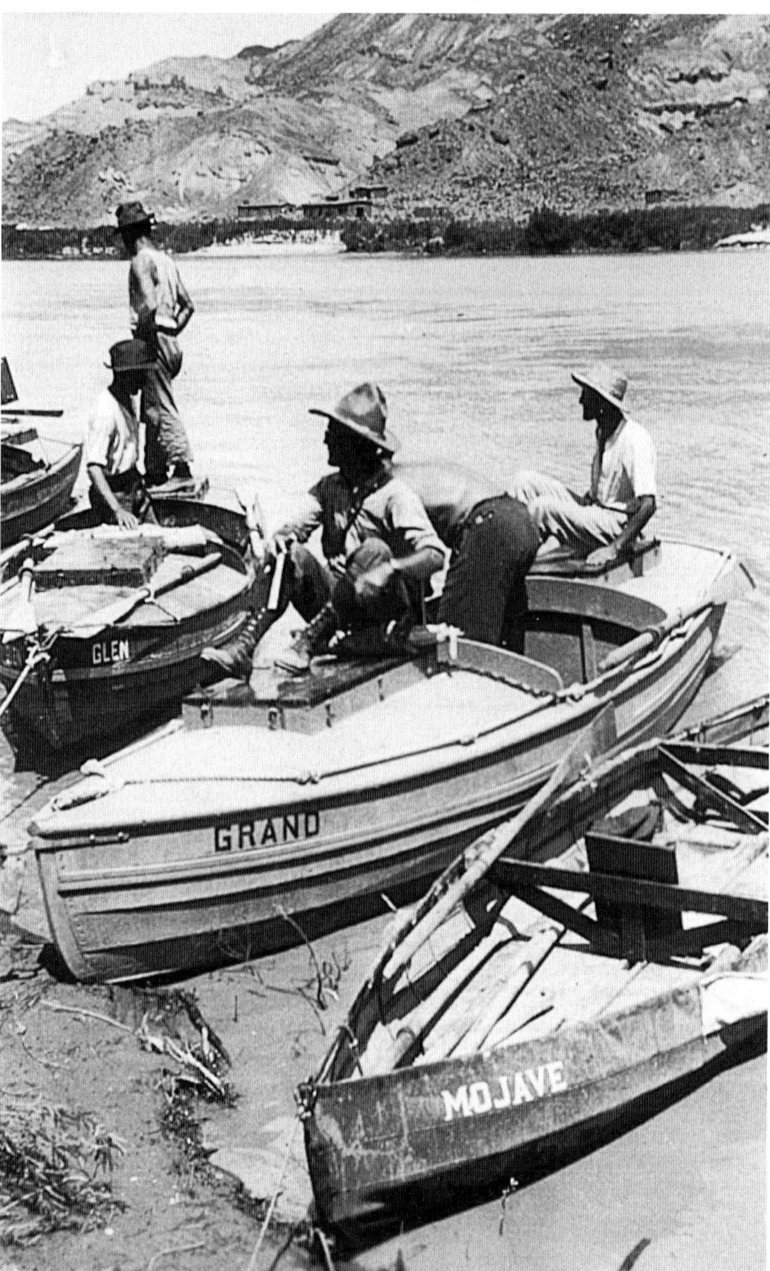

R. C. Moore aboard the Grand *as the 1923 crew prepares to shove off at Lees Ferry (R. C. Moore Collection).*

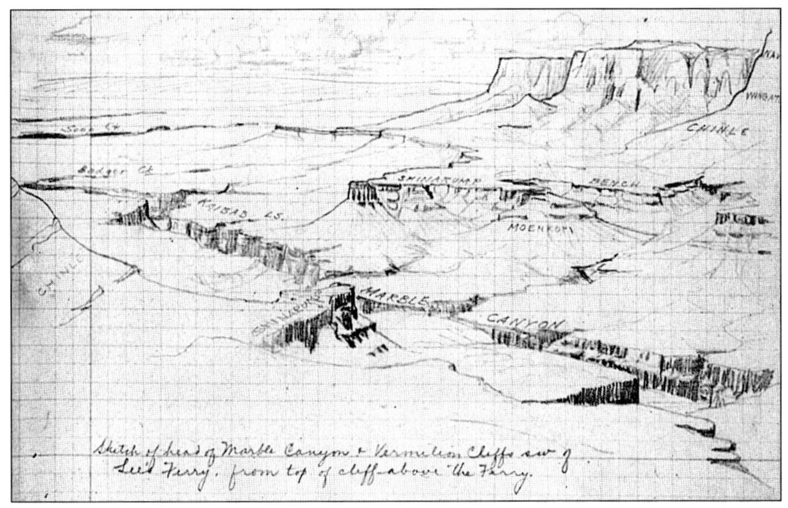

Moore's sketch of the head of Marble Canyon, from southwest of Lees Ferry. Moore was an accomplished artist; he drew detailed cross sections and later sketched numerous portraits of friends and colleagues (from Moore's Grand Canyon notebook, R. C. Moore Collection).

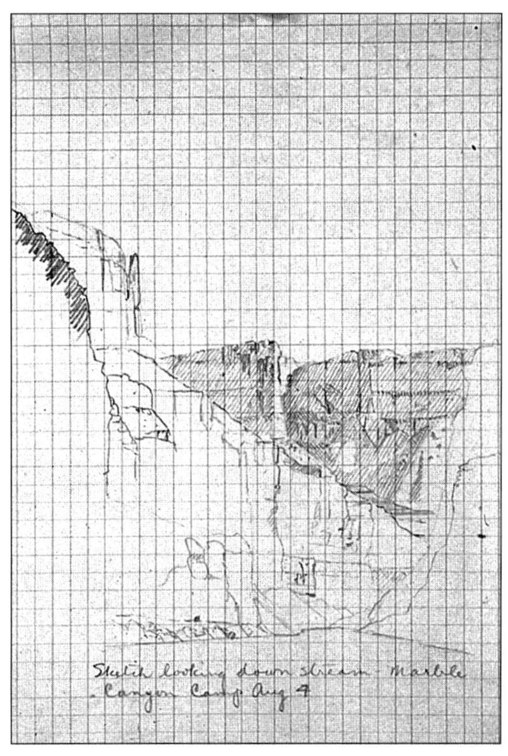

Moore's sketch looking down Marble Canyon from the mouth of Badger Creek (from Moore's Grand Canyon notebook, R. C. Moore Collection).

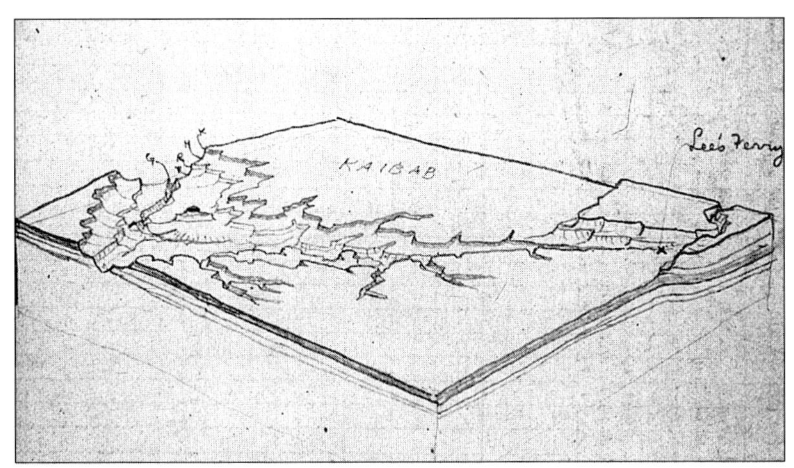

Moore's block diagram of Marble Canyon (from Moore's Grand Canyon notebook, R. C. Moore Collection).

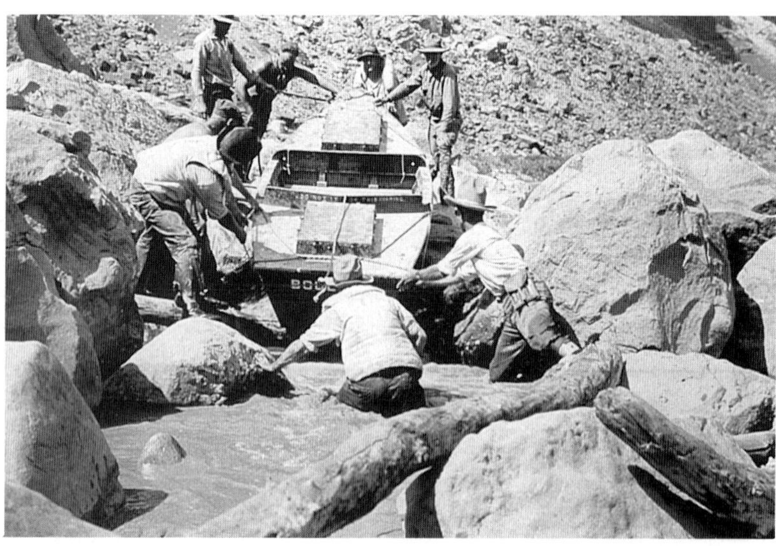

The 1923 crew portages the Boulder *around much of Soap Creek Rapid (R. C. Moore Collection).*

would work here, several hundred feet below the canyon rim—and heard the news that President Harding had died. Our group gathers for supper, followed by an introduction to the protocol of life on the river. The Hatch crew gives instructions about bathing (soap and shampoo are only allowed in the main channel of the Colorado, not in the side streams), about washing our dishes (the final rinse is in hot, chlorinated water; nobody wants a case of dysentery down here), and about picking up after ourselves (everything that comes into the canyon must go out).

The 1923 trip was well documented. Many crew members kept detailed journals, describing both the canyon and their experiences. The most widely read account was probably that of Lewis R. Freeman, a boatman who wrote a long article about the expedition for *National Geographic* and then expanded the story into a full-length book called *Down the Grand Canyon.* Emery Kolb decided that Soap Creek Rapid, just below our camp, was too dangerous to be run and would have to be portaged; Freeman later wrote, "There was some disappointment at missing the chance to attempt what might have been the first successful run of this notorious rapid, but there would have been no point in taking undue risks with an outfit like ours, especially so early in the voyage."

SEPTEMBER 9, 1991: Our day at Soap Creek starts clear and cool, the sun hidden for much of the morning behind the canyon's east wall. Photographer John Charlton takes three photos at Soap Creek, one from above the river (photo pair 4) and two at river's edge (photo pairs 5 and 6). The photographing lasts well into the afternoon. Later we try to find another photo taken by the 1923 crew a few miles below Soap Creek, but we can't locate the photo's vantage point. Without an identifiable feature to focus on—like a rapid or a particularly distinctive rock formation—it is difficult to pinpoint any single location on the river.

Many of the original photographs were either not labeled or

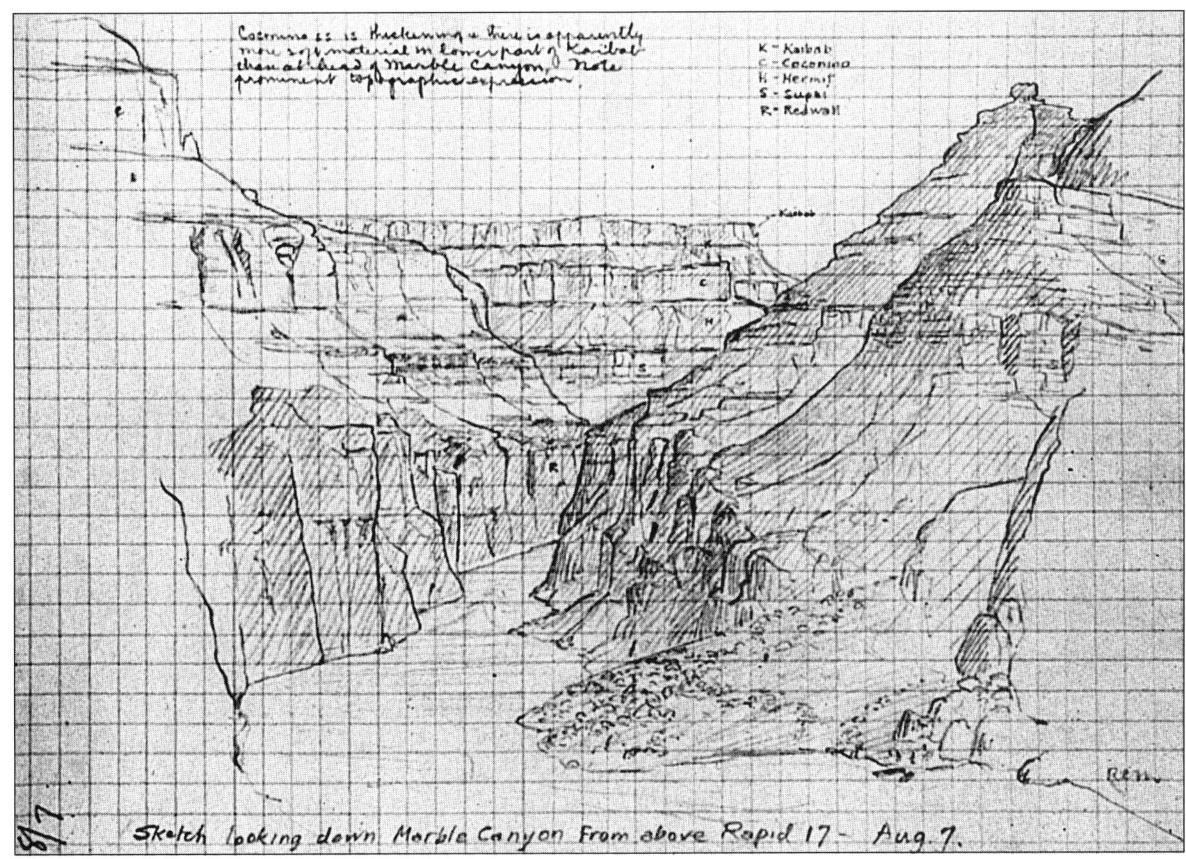

Moore's sketch looking down Marble Canyon from above Rapid 17 (from Moore's Grand Canyon notebook, R. C. Moore Collection).

wrongly labeled. That seems odd to me. The 1923 crew appreciated the historic nature of their work, but didn't they realize that the photographs they took were more than snapshots, that they would enter an archive of canyon history? Relocating these photographs requires an intimate knowledge of the canyon that could only be gained by repeated trips down the river.

We do find a photo location above Sheerwall Rapid (photo pair 7). The place gets its name because sandstone walls rise almost straight up from the river and provide perches from which one can see the boiling eddies in the brown water. At about this spot, 1923 boatman Elwyn Blake wrote, "Some of the swirls and eddys are so strong that they seem capable of drawing the stern of a boat completely under. Rowing was very difficult."

All this water looks somehow out of place cutting through the midst of aridity—through what is essentially a desert. The river even has the characteristic muddy, floodwater smell of large bodies of water. A float trip down the canyon is obviously

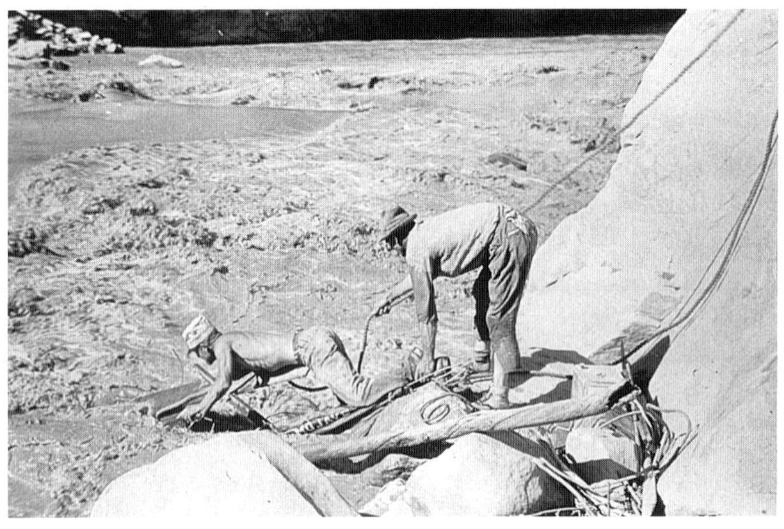

The wreck of the Mojave *(R. C. Moore Collection).*

different from a visit to the rim. Down here, the river is central, integral to the canyon. From the rim, the river seems incidental. From a few spots on the South Rim, though, you can hear the rapids roar seven miles away. The river at work.

We stop for the night at a crowded cluster of rocks and sand on the west side of the river at about mile 19. The 1923 crew probably stopped about here. Blake said that "camp was pitched on a rocky shore. Luckily there was a sandy bench near by, where the beds could be unrolled." We spend the night at a similar crowded, rock-strewn camp, where I take my first river bath of the trip. Quickly. The water coming out of Lake Powell measures 40 degrees Fahrenheit.

SEPTEMBER 10, 1991: This morning starts with a long run down a gullet of canyon walls formed by Redwall Limestone, a stretch of canyon so spectacular that it sometimes stops conversation. In places, the red-stained walls of the canyon are coated black with desert varnish; in other places, rainwater has cascaded down the canyon wall for decades, polishing it a shiny, sparkling white that looks almost like snow.

We float past a rapid where the 1923 crew lost its canvas boat when it was caught on a submerged rock. According to boatman Blake, "the current immediately tore the floor boards out and swept them away. Later efforts to pull the craft free only succeeded in tearing [it] to pieces." The wooden boats allowed room for only one or two people to sit and row, while the other passengers rode on the boats' planking, holding onto lines stretched across the decks. Going through one rapid, Blake wrote, "geologist Moore rode on the *Boulder,* lying face down on the stern hatch. Several waves broke over the stern of the boat, drenching Mr. Moore thoroughly."

Riding through rapids today is more sedate, although passengers occasionally straddle the front of the boat's pontoons. Called "riding the horns," this is a good way to get wet, cool off, and feel the river's power at the same time. The big rapids throw up walls of water that hit with the force of a football player.

Our first photo today is at South Canyon, just above Vasey's Paradise (photo pair 9). We hike up a trail to a point about seventy-five feet above the river, and on the way Bill Ellwanger, our boatman, shows us an insect web on a prickly-pear cactus. He carefully plucks the white web off the cactus spines and squashes it between his fingers, smashing some insect larvae inside and turning the web blood red. The Navajos, Bill says, use the insect coloring as a dye in making blankets.

Our next stop is for two photos at the base of Vasey's Paradise (photo pairs 10 and 11), where warm, clear water flows from waterfalls, through vegetation, into the river. Then we move to Redwall Cavern for two more shots (photo pairs 12 and 13). Each photograph requires an hour or two, so we leave the cavern late in the afternoon. We stop briefly to take another photograph (photo pair 14) as the evening's light starts to fade.

This is about the same point where the 1923 crew studied the geology for several possible dam locations. One problem, they found, was that the Redwall Limestone is prone to dissolution and the formation of caves like Redwall Cavern. This makes the Redwall a poor dam foundation. In his notes about the trip, R. C. Moore wrote: "The Redwall is cavernous and traversed by many underground waterways in and just below the fault zone, but above, in vicinity of damsite, the limestone is massive and solid." In the 1960s, builders proposed to construct a Marble Canyon dam at about mile 40. As we float past, we can still see drill holes from engineering tests in the canyon walls, a tangible reminder that people once intended to build a dam here, drowning everything we've seen so far.

By now we are running late. On our way to camp, the canyon colors change with the dwindling light, from bright reds to dull, glowing pinks, finally giving way to shades of gray as the light disappears altogether. As it grows darker, the river becomes more dangerous—boulders are almost impossible to see, which means that Bill is steering a course from memory—and the fun fades with the light. Finally, we see a lantern on the shore and pull over at the night's camp next to President Harding Rapid.

September 11, 1991: The 1923 crew laid over here on August 10, 1923, the day of Harding's funeral, and then named the site after the president. They washed clothes, listened to the radio. Our morning is taken up with photography (photo pairs 15 and 16), then a move to Nankoweap Rapid.

In 1923, Kolb and Blake hiked up a trail to a spot about six hundred feet above Nankoweap Rapid, where the Anasazi had built a series of granaries into the canyon wall to store their seed corn, ears so small that they're now called "middle finger corn." We make the same steep climb. The view is spectacular but the drop so steep that all I can do is cling to the canyon wall. After twenty or thirty minutes, my vertigo passes, and I manage to climb up and into one of the small granaries. It's somehow unnerving to realize the Anasazi sat and worked in this shadowed stillness a thousand years before.

We've spent much of this trip following the paths of others, sometimes the Anasazi, sometimes Powell, usually the 1923 crew. As we line up the rephotographic shots, our view and the old photo sometimes match up precisely, everything clicks into place, and we know—*know*—that we're standing exactly where one of them had stood. Matching the photos is at first frustrating, then thrilling, and last a little unsettling, almost as if the ghosts of those 1923 explorers are nodding when we finally get the location right. It makes me wonder whether someone might follow us in another sixty or seventy years, trying to duplicate our photos. What will they think? Will they grapple with the same environmental dilemmas that we face? In some ways, it is easier to envision the mysterious Anasazi than to imagine people of the future.

Time has eaten at the canyon wall in the years since 1923. The vantage point they used for the Nankoweap photo was a ledge of rotten shale, much of which has fallen away in the meantime. Bill and John clamber out onto what's left of the ledge to take the photo (photo pair 17). From here, we can see far down the canyon. Sun lights up the opposite wall. At times the light is interrupted by clouds that transmute the reds to moving spots of gray. From up here, the river is brownish blue, lined by bright green vegetation. Everything is color.

We go back down the trail for supper. As the evening passes, more boats pull into Nankoweap, turning the river into a watery parking lot. Newcomers stumble into tents in the dark. On the beach, a television crew films an interview with a canyon biologist. The camera crew runs a generator that rattles; their lights brighten the beach and attract bats. About twenty thousand people go down the river each year, but the canyon seems to bear up, its camps kept clean by stringent regulations and the environmental consciousness of the river runners. The canyon is obviously under incredible pressure.

September 12, 1991: This morning's first shot (photo pair 18) is above Kwagunt Rapid, in the midst of tamarisks, or salt cedars, feather-leafed trees that have colonized the beaches along the rivers. Most of our photographs show little change in the geology, but nearly every one shows far more vegetation, often in the form of tamarisks, than was present in 1923. Glen Canyon Dam evens out flows in the river, allowing the vegetation to establish itself on the beaches to the point that it's a nuisance.

Next stop is the confluence of the Colorado and the Little Colorado (or Little C, as Bill calls it). The first sixty or so miles of the trip—from Lees Ferry to the confluence of the Colorado with the Little Colorado—is through Marble Canyon. Here, at last, we officially enter the Grand Canyon. The Colorado had been clearing up during our trip, but the Little C muddies it again. Don Baars, who has been down the river maybe thirty times, says that the Little Colorado is sometimes as much as 80 percent silt, more a mudflow than a river. In 1923, Lewis Freeman wrote that it was "an unseasonably high stream of vile-smelling mud." Sometimes it's azure from calcite dissolved in the water. Not during our visit. We hike up the Little Colorado to repeat a couple of photographs (photo pairs 19 and 20) from flat benches of coarse Tapeats Sandstone.

At about mile 63, as we slow down to look at salt that has seeped out of the canyon wall, Don Baars spots a hollow in the Tapeats Sandstone on the opposite wall where the 1923 group photographed a crew member reclining in the rock (photo pair 21). We hadn't expected to find this spot; the river is lined by miles of Tapeats, much of it pocked with holes that give it a Swiss-cheese appearance. Finding one irregular hole in the rock is like looking for a particular penny in a jar full. But Don has found it.

Tonight we camp just across the river from the Great Unconformity, where rocks display the gap in geologic time between the Cambrian and Precambrian eras. I can't help contemplating the Grand Canyon's hold on nearly everyone. Its allure for geologists is understandable; the canyon displays an unparalleled geologic cross section, along with other highlights—miles of lava flows, delicate travertine deposits, massive faults, unconformities like the one across the river. But it's more than that. The canyon's attraction almost certainly has something to do with the Southwest. If the canyon were in, say, New Jersey, would it draw the same crowds? The desert environment, the ghosts of the Anasazi, and the presence today of Native Americans—are they what bring people here? Or is it the obvious age, the huge expanse of time the canyon represents? This rock gives us a glimpse of the near-eternal, and set against it, our lives seem small. Is it this paradox that teases our thoughts?

SEPTEMBER 13, 1991: This morning begins with a quick hike across Chuar Flats to look at an old mine works—equipment cast aside and rusting, a hole in the high wall now dripping with water—then a long float to Nevills Rapid for a photo (photo pair 22). The next photo (photo pair 23) is from a ledge of thin-bedded, black Precambrian sandstone about one hundred feet above the river, where it finishes a right turn and changes course from generally south to generally west.

We stop just beyond this bend at Hance Rapid, a long, rock-choked rapid rated a tough seven on the river runner's scale. The 1923 crew got here on August 16, a little ahead of schedule, because much of this area had already been surveyed and studied by workers who came down trails along the Little Colorado. Kolb, Blake, and Leigh Lint hiked up Hance Trail to the South Rim, where they spent the night, went to the Kolb photography studio, and celebrated Lint's birthday. Then they redescended into the canyon with a pack train to resupply Birdseye's group.

We don't have time for hiking. It gets dark early, about seven o'clock, and people go to bed a couple of hours after sundown. There is only a little moon on this night, and in the clear, dark sky the stars come out in such profusion that they seemed draped across the skies. On my back in a sleeping bag, I count three shooting stars before going to sleep.

SEPTEMBER 14, 1991: We spend the morning taking photos (photo pairs 25, 26, and 27) at Hance Rapid. With the morning sun behind the canyon rim, the far walls are gray, the near walls a dull red. One thick, dark gray dike cutting through the red rock is paralleled by an inch-thick white dike. The walls behind us are a soft brown. I'd like to think that the Anasazi had names for each hue, the way the Eskimos supposedly have names for umpteen varieties of snow.

We finish the photography, then load up the boats to run the rapids. This puts us in Granite Gorge, with its polished black Precambrian rocks like the Vishnu Schist: surely one of the most melodic geologic names ever. Here the Vishnu looks as pretty as it sounds—the black surface worn fluted and smooth by the river. Our run through Sockdolager and Grapevine rapids is without event. The 1923 group ran these rapids too, because the walls here are so sheer that there's no way to portage around rough water. Just after Grapevine, we stop for a photo (photo pair 28) from a small talus pile of Vishnu Schist, its surface hot in the midday sun. After the photo stop, we eat lunch on the boat—sandwiches and lemonade, with the river and high walls of polished Vishnu for a backdrop.

We move on, floating under the bridge that's part of the Kaibab Trail. We pick up water at Clear Creek. Pass Phantom Ranch. Float beneath the bridge for Bright Angel Trail. See a few hikers. The canyon furnishes, by turns, isolation and crowds. On one hand, it's hard to get out of the canyon, except by helicopter or a long hike at places where there's a trail. There are no newspapers or telephones, except at Phantom Ranch. In this expanse of canyon, it's easy to be alone. On the other hand, we see other boats on the river and at the campgrounds almost constantly. And this is September, the end of the tourist season. What is it like in July or August?

The 1923 group tied up here for a couple of days after leaving Hance Rapid, and several members trooped up Bright Angel Trail to the South Rim. They gathered their mail and took a tour of the sites; two of them even went to a dance one night. They had been in the canyon just over three weeks. We have just finished our first week.

We take a photo just below Bright Angel Trail (photo pair 29), then stop for the night at Granite Rapid. After dark, a ringtail cat scampers through camp.

September 15, 1991: Our first shot of the day is at Hermit Rapid (photo pair 30). A second photo here is impossible to replicate because of the growth of tamarisks. The 1923 crew took movies of themselves running the rapids here. Footage from that trip shows the boats bobbing and bucking, lifted up and slapped down by the muddy waves of the river. Shots filmed onshore seem slightly comical, the crew walking with the herky-jerky acceleration of silent movies.

We make the run through Crystal Rapid, then move on to set up camp near Bass Rapid. The 1923 crew stopped here too, waiting for supplies to come down Bass Trail. This system of packing supplies down from the rim allowed the crew to stay in the canyon for an extended period. We spend the afternoon shooting a photo just above Shinumo Creek, looking back upstream. Because people travel down the river, they generally look ahead; I would have guessed that most of the 1923 photos would depict the view downriver. Instead, many of them show what the crew had just passed (maybe to depict possible dam sites) and these backward-looking photos are difficult to match.

We take the photo at Shinumo Creek (photo pair 31) in the hot sunshine of the afternoon, then head back to camp. As the sun goes down, we watch as a butte of Redwall Limestone disappears into the shade. Evening shadows move up the side of the canyon, working their way through the geologic cross section and geologic time, first covering the lowest, oldest rocks, the Precambrian, then the slightly younger Cambrian and Devonian rocks, until finally reaching the 350-million-year-old Redwall and casting it into darkness.

September 16, 1991: Our first photo of this day, a Monday, is at Waltenberg Rapid (photo pair 32). The 1923 crew portaged this one. On our visit, the water levels are low, exposing rock, because of lowered releases from Glen Canyon Dam. The dam generates electricity for Phoenix, Los Angeles, and other Southwestern cities. When they don't need as much juice—on Sundays when businesses are closed, for example—less water is released, making less electricity. It takes a little more than a day for water to get from the dam to here, about a hundred miles downstream, so the lessened flow takes a while to show up.

Flow rates are a contentious subject. At one time, the dam operators could release any amount of water they wanted. The resulting changes in flow, especially when lots of water was released suddenly, were tough on the canyon, tearing up sandbars and increasing erosion. Now releases are regulated, and changes must be made gradually. Still, there's something disconcerting about the way the water level bounces up and down. It's like a river with a tide.

The day's second shot (photo pair 33), from a perch about a hundred feet above the river, demands a scramble up a steep slope of loose schist and sliding talus. Hanging out over the river, we can see the far rim capped by the dusty Tapeats Sandstone, and below that a sheer wall of shiny Vishnu Schist shot through with lines of sharp red Zoroaster Granite, all of it dotted green by barrel cactus and huge prickly pear. It isn't hard to see why this site attracted the 1923 photographer. I wonder if anybody has climbed up here since.

We shoot another photo below Waltenberg Rapid (photo pair 34), make a quick stop at the delicate, travertine-covered rocks and waterfalls of Elves Chasm, then head on to camp at mile 119. There are fewer photos to duplicate the rest of the way. The 1923 crew simply took more photos at the beginning than they did later in the trip—typical of most people who go down the river, says Don. As people go down the river, they become accustomed to the scenery and take fewer photos. They become numb to the spectacular.

The Colorado River is like a series of steps: long stretches of quiet water followed by steep drops at the rapids. Camp tonight is at a quiet place. Without the noise of rapids to drown out snoring, things get pretty loud. I have strange dreams, unsettling and vivid. In one, a friend gets divorced, in another Don Baars has a new job selling used cars on TV. Maybe I should study sleep disorders instead of geology.

> Before the children of Europe took these hills, the people who walked here believed stones to be alive because they carried heat, changed their forms, and moved if you watched long enough.
>
> —William Least Heat-Moon
>
> *Prairyerth*

September 17, 1991: We float Conquistador Aisle, the river again lined by the hole-riddled Tapeats Sandstone, then move into middle Granite Gorge, another stretch of the Vishnu Schist. We match one photo location from a talus pile on the right side of the river, a second from another precarious perch on the left side, looking back at 128-Mile Rapid.

These two photos typify the change and absence of change that we have seen throughout the trip. The first photograph (photo pair 35), taken near the water's edge, shows virtually no change. Down here, where the Vishnu forms the canyon walls, the schist is hard: it wards off erosion and is too tough to support much vegetation. Even the almost-omnipresent tamarisks find no purchase. It's startling that in sixty-eight years, with this powerful river coursing along, sometimes in flood stage, there has been so little change. It provides a new perspective on geologic time. At this spot, sixty-eight years isn't the blink of an eye—it's no time at all.

But time has been at work at 128-Mile Rapid (photo pair 36), both at the rapid's edge and in the higher walls above the river. Erosion has changed the canyon walls, and chunks of rock have spalled away from their original location.

The fact is, most of the changes we spot along the river are the result of human activities. Many, such as vegetative change or beach erosion, are related to the building of Glen Canyon Dam. In only a few instances, as on the far wall here or in places where floods have rearranged the rapids, are changes noticeable and obviously natural.

September 18, 1991: This is our only day off from photography, and we spend it hiking up the

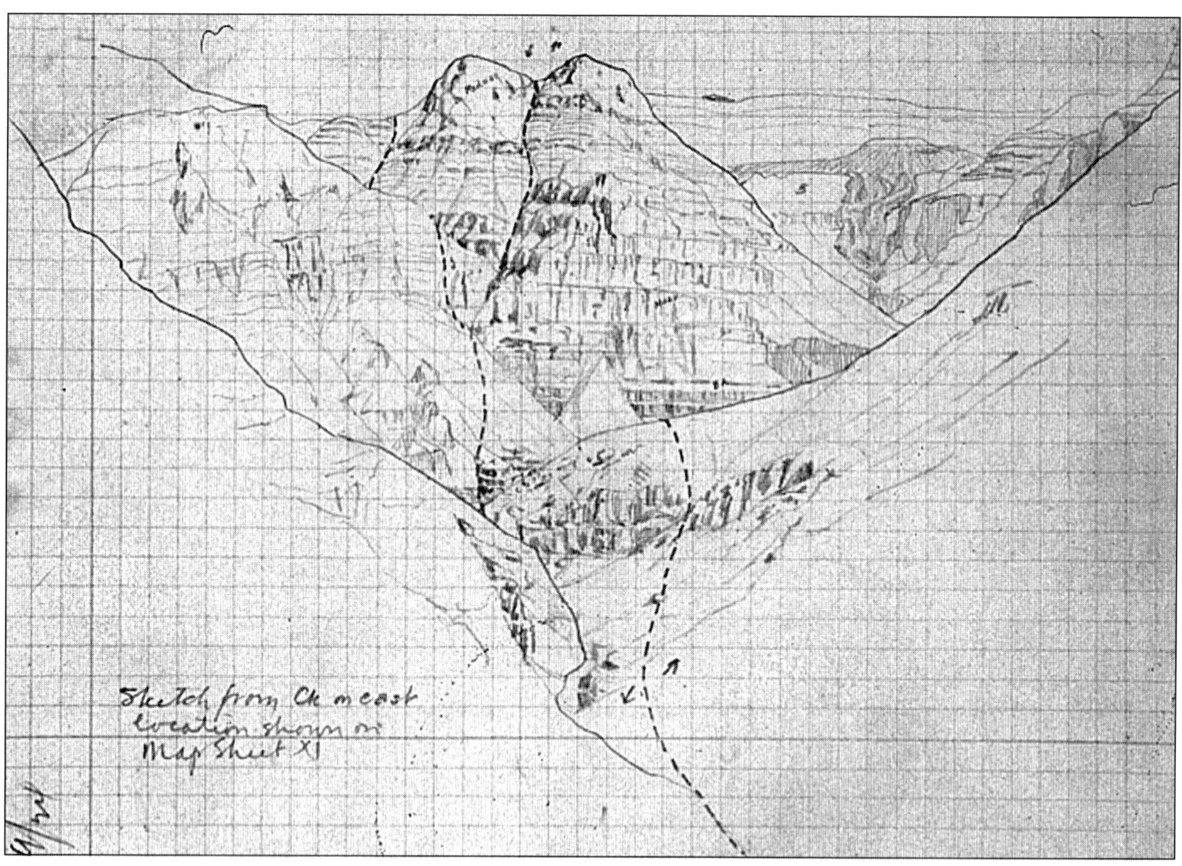

Moore's cross section at mile 149.8 (from Moore's Grand Canyon notebook, R. C. Moore Collection).

side canyon at Tapeats Creek. The path runs up the canyon wall, through a cactus-dotted valley, and over three creek crossings to Thunder Falls, a place where water explodes from fissures in a steep slab of Redwall and then slides into a grove of cottonwoods and willows, watercress and mint. We eat lunch in the mist of the falls while I read Blake's account of the same hike in 1923: "Huge prickly pear were in bloom. Great cholla cactus studded the grass covered slopes, while the stream plunged from rock to pool in white cascades. The shade of cottonwood trees was welcome, too. About noon we reached the forks of the stream. Here we came suddenly upon a fountain of water gushing from a fern and moss covered bank. It was a lovely sight." Some things don't change.

On the way back to the boat—a long, hot hike—we run into a pack of British tourists. When we get back to the Colorado, five or six big rafts are tied up at river's edge and a dozen kayakers are launching into Tapeats Creek Rapid, their boats skittering over the waves like water bugs. The 1923 crew was virtually alone on the river. Some things do change.

Word has it that George Bush is on the South Rim today for

some kind of photo op. Couldn't tell it from down here.

September 19, 1991: The first stop of the day is at Deer Creek Falls (photo pair 37), followed by a hike up Kanab Creek, a polluted little stream that dumps into the Colorado from the north. We hike through a slick, muddy creek bed, its edges coated with a crust of white salt. This steep-walled side canyon doesn't get hiked much, although there are footprints even here. In 1923, the creek was clear and the crew hiked up through the water, took baths, and shaved. One of the photos that we try to relocate shows a boatman bathing in Kanab Creek, along the edge of a level bench of Bright Angel Shale. After a two-mile hike up the canyon, we give up. Maybe we didn't hike far enough, or maybe we walked past the original location and didn't see it. Maybe it's changed beyond recognition.

Back on the Colorado, about seven miles below Kanab Creek, the 1923 crew upset the boat *Marble* at an unnamed rapid. The boat flipped over and Emery Kolb, who was rowing, was tossed out. Blake, watching from shore, described it this way: "The *Marble* was caught by the back curling wave below. In the next instant she was upside down, bounding swiftly through the tail of the rapid. Emery was nowhere to be seen . . . and then we saw Emery's white hat emerge. Fortunately, Emery was under the hat." The crew called this spot Upset Rapids, a name that it bears today.

September 20, 1991: Havasu Creek may be one of the most photographed locations in the Grand Canyon. The creek enters the Colorado from the south, running through a narrow canyon lined with delicate white travertine, the water a light green, the color of oxidized copper. This is the only location on our trip where we duplicate a photograph from the boat (photo pair 39); we park right in the mouth of Havasu Creek, which is probably where the 1923 crew took its picture. When we stop at Havasu, several boats are tied up at river's edge, including the boat of the British tourists, who have dutifully trooped up the trail to see Havasu Falls. But the creek is deserted and we have it to ourselves for a glorious hour of swimming and sunning.

This stretch of the canyon, above and below Havasu Creek, is formed by steep cliffs of Redwall Limestone, highly sculptured with smooth faces where side streams and waterfalls come in. Birdseye and Moore wrote later that "the massive limestones are brought to the water's edge, and great, nearly sheer cliffs rise 1600 feet on each side, with benches up to 2500 feet farther back. This is the deepest narrowly enclosed gorge that was found in the journey through the canyon." Below Havasu we pass a field of hoodoos—tall rock pillars—called Red Slide, and large lava flows with rock the color of the bottom of a burned chocolate-chip cookie.

We stop for a photo (photo pair 40) just above National Canyon, at about mile 166. A light pink Grand Canyon rattlesnake—the first we've seen—is occupying John's vantage point and delays shooting for a time. We camp this night just above Lava Falls. The moon, now almost full, lights up the canyon like a night scene in a cheap western.

September 21, 1991: Lava Falls is notorious. The rapid's thirty-seven-foot drop, rated a ten, makes it one of the most difficult rapids on the river. Even the big boats respect Lava Falls, and the oar-powered boats almost always unload, their occupants scouting the rapid before they proceed.

Lava Falls was even more troublesome for the 1923 crew, although for a different reason. They arrived here on September 18, unloaded most of the gear from the boats, and then lined the boats through the rapid and camped below. During the night, water from a flash flood in the drainage basin of the Little Colorado reached Lava Falls. The river rose twenty feet. Warnings were broadcast to the crew over the radio, but their receiver had failed earlier in the trip and was hauled out at Havasu Creek to be repaired and returned later. Thus the flood caught them unprepared, and they spent most of the night continually moving the boats to higher ground so the river wouldn't smash them. Freeman wrote that the flood water had "a distinctly unpleasant odor. . . . Mingled with this was a humid, almost tropical, smell—rank vegetation and a suggestion of the perfume of flowers."

For three days the crew waited for the water to subside. The delay made them late to Diamond Creek, their next resupply point, and that led to fears that they might have perished in the flood. The student newspaper at the University of Kansas ran a story with the headline "Exploration Party May Be Lost." The crew killed time at

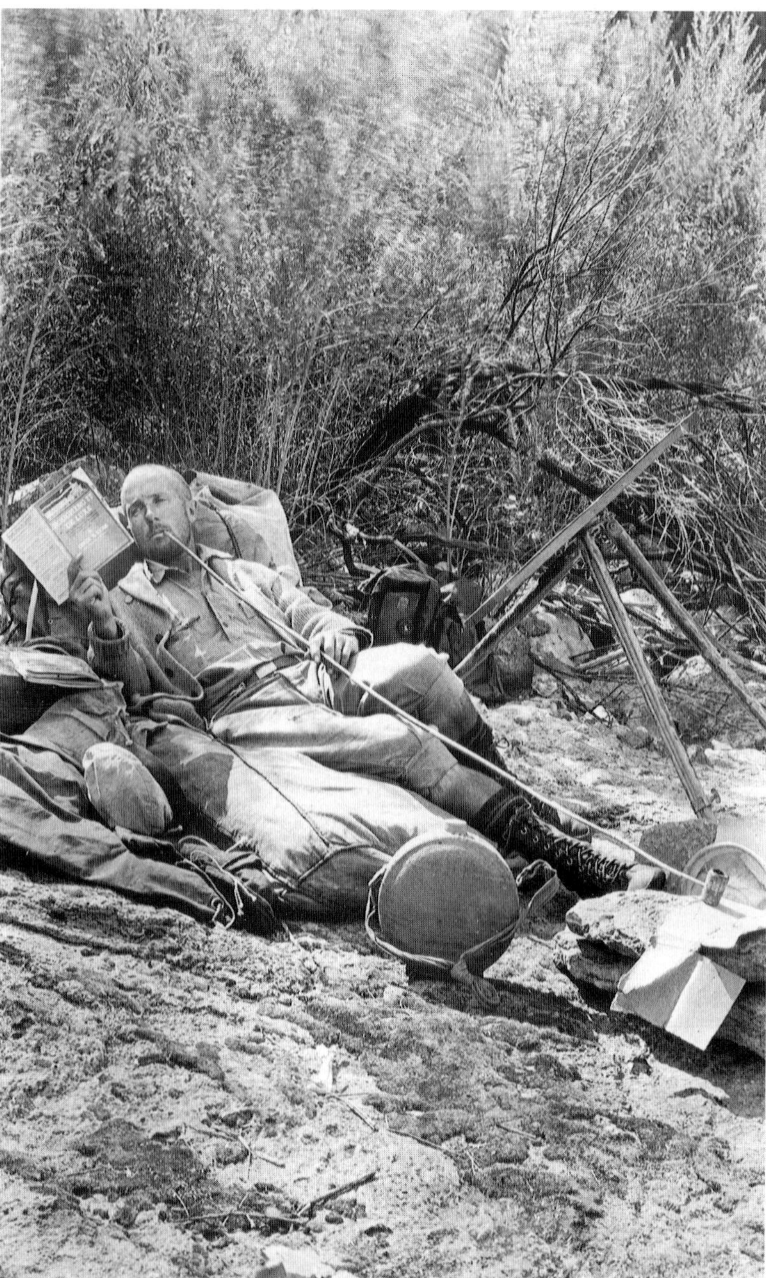

R.C. Moore relaxes while waiting for water to recede at Lava Falls. The book in his hand is titled Roughing It Deluxe *(E. C. LaRue, USGS).*

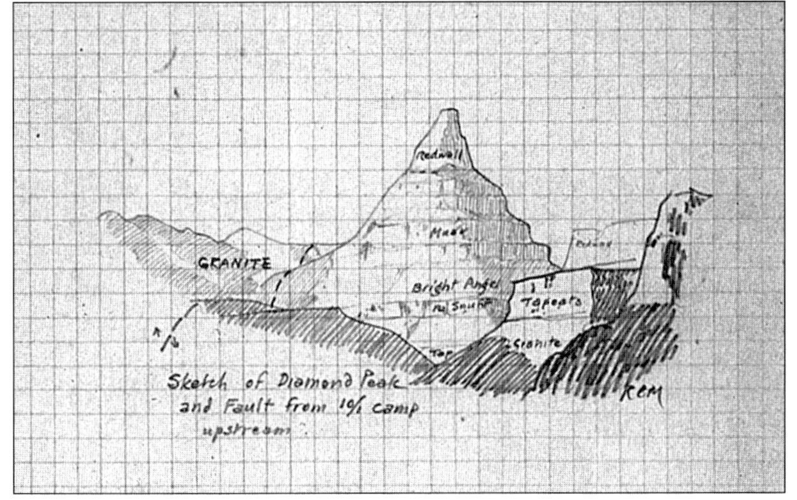

Moore's sketch of Diamond Peak (from Moore's Grand Canyon notebook, R. C. Moore Collection).

Lava Falls by taking care of the boats, hiking, and watching the monstrous waves of the flooded river.

Their last day at Lava Falls was September 21, 1923, exactly sixty-eight years to the day before we floated through the same rapid. Their photos, with the Colorado flooding, make Lava Falls look so different that it's hard to imagine we're at the same spot on the same river (see photo pair 42). Our run through Lava is without event, though several standing waves almost submerge our boat. Afterwards, we wash off in a warm travertine spring.

> Nothing lasts. Old seas are buried. A species of dinosaur may go extinct after only six million years. We will live to be eighty, with health failing after sixty: a fact.
>
> —Rick Bass
> *Oil Notes*

September 22, 1991: Our last stop is just above Diamond Peak, where we shoot three photographs (photo pairs 43, 44, and 45). The climb to the final photo is through the usual assortment of cactus—prickly pear, barrel, beavertail, fishhook. While we're shooting the photo, a thunderstorm kicks up wind and sand; Bill grabs the tripod to keep it from blowing away. On the way back to the boat, John flushes out a second rattler. It's as if the canyon wants us out.

In the evening, a slow drizzle settles over our last camp. This night, on the verge of going home, I remember the first time I was away from home for a week. Church camp. I was ten years old. When I got home, I expected everything to be different. Changed. But it wasn't. Everything was the same.

It makes me think again about the canyon's future, how things might look here in another sixty-eight years. The geology? Our photos show little change in the canyon's geological features over the past seven decades; barring some geological catastrophe, it won't change significantly in the next seven, except maybe around the rapids or along the beaches. The vegetation? More changes will probably occur, as more growth takes hold. And the people? Will they row through in oar-powered boats, having run out of fossil fuels? Will they still be wrestling with global warming and overpopulation, the legacy of our profligacy? Will there be even more restrictions on visiting the canyon, made necessary by people's search for natural experiences in a crowded, human-dominated landscape?

If you graphed the growth of environmental concern in the twentieth century, you'd see big spikes in the early 1970s and the late 1980s. What will the people of 2059 think about their environment? Will trips down the river still be a treasured encounter or will environmentalism be viewed as some passé Romantic movement? All I know for sure, on this final night of the trip, is that in sixty-eight years the people now settled comfortably on this beach will be meat for worms. Our words and photos may or may not exist. But the canyon, the canyon will abide.

SEPTEMBER 23, 1991: We get out of the river at Diamond Creek, about mile 225. The 1923 trip went on to Needles, finally leaving the river on October 19, 1923. Their adventures continued. At Diamond Creek Rapid, Freeman's boat smashed into a rock, opening up a six-inch hole. At Separation Rapid—where three members of the 1869 crew left Powell, climbed out of the canyon, and met death before they reached a town—the 1923 boat *Grand* was turned over and its passengers pitched into the water. Moore was able to climb on board, then grab Freeman and pull him back into the boat. The only losses were Moore's sunglasses, several hats, a canteen, and two geologist's picks.

We drive out at Diamond Creek, through the little town of Peach Springs, stopping in Flagstaff for lunch. Japanese food. Pee Wee Herman is on the cover of the *Rolling Stone* at a nearby bookstore. Back at Marble Canyon for the night, we take one last photograph, this one from the rim, with Navajo Bridge in the background (photo pair 1). While we work on the photo, the sun drops, Echo Cliffs glow red, and we listen to the question-and-answer call of birds in the evening. We check into a motel to shower and watch Monday night football on television.

After their trip, the 1923 crew took rooms in a Needles hotel. Surveyor Frank Dodge and boatman Blake went to a theater that happened to be showing a Fox newsreel of their trip. "Dodge and I went to the movies," wrote Blake. "We were stopped cold in the aisle as we entered. There on the screen was the *Glen*. I was at the oars and Dodge lay flat on the after hatch. . . . 'That's us, in Hermit Creek rapid,' I whispered."

An Open Book

Donald L. Baars

Grand Canyon, Arizona: arguably the greatest natural geological laboratory on earth. There the Colorado River, cutting a chasm through the earth's crust for 280 miles, in places as much as a mile deep, exposes eons of earth history in magnificent detail. John Wesley Powell, the first river runner and a self-taught geologist, wrote: "All about me are interesting geological records. The book is open, and I can read as I run."

The Colorado River rises near Long's Peak in the high country of Rocky Mountain National Park, Colorado. It doubles in volume and muscle at its confluence with the Green River in the heart of Canyonlands National Park in Utah. Downstream, the river is first crippled by Glen Canyon Dam and then reborn as it plunges into Marble Canyon at Lees Ferry, Arizona. There the river drops abruptly into the depths of canyons carved into rocks of Paleozoic age, formed between 500 and 260 million years ago.

It is common knowledge that rivers cut their canyons like conveyor belts carrying silt and sand that relentlessly grind the river bottom throughout millions of years. But this process only deepens the canyon; it has nothing to do with widening it. What broadens canyons is the endless repetition of little events: flash floods, rock slides, rockfalls, mudflows, wind erosion, the footsteps of animals. Soft rocks in the canyon walls, such as shale, erode to form smooth, relatively gentle slopes. Hard rocks, such as the sandstones and limestones that are most affected by rockfalls, form near-vertical cliffs. Alternating layers of hard and soft rock yield stairlike sequences of slopes and cliffs.

Why did the Colorado River choose this site to etch its deepest fissure into the Colorado Plateau? No one knows for sure, although there have been many educated guesses. There is, however, little doubt that the great rivers of the Southwest achieved roughly their present courses by middle Tertiary times, some 30 million years ago. Then, the landscape we know today was buried under at least five thousand feet of sedimentary rock that has since been washed away to distant settling ponds—most went to the Pacific Ocean via the Gulf of California. Then, too, the land was flat, lying near sea level. With little gradient, rivers meandered aimlessly.

But a time came when the entire western North American continent began gradually to be uplifted. Rocks that had formed as shallow-water sediments more than 250 million years ago slowly rose to heights of one or two miles above present-day sea level. The rivers began to cut downward, trapping themselves in valleys and canyons from which they could not escape. As they encountered older and harder rocks, rivers that lacked sufficient power to cut through the rock became impounded as great lakes until the ever-steepening gradient caused the waters to overflow the hard-rock dams and continue toward the sea. In this way, canyons like the Grand were chiseled deep into the hardest rocks on earth.

Erosion tries with all its might to level the land. Where rocks have been uplifted, as on the Colorado Plateau, the younger, or uppermost, rocks are stripped away first, exposing the more ancient rocks beneath. It is just this process that allows the Colorado River, beginning at Lees Ferry, to expose the oldest rocks of the Southwest in the depths of the Grand Canyon.

At Lees Ferry, a long, asymmetrical upfold known as the Echo Cliffs monocline brings older rocks up to the surface of the land to form what is called the Marble Platform. Hard, resistant Kaibab Limestone dating from the Middle Permian, or about 260 million years ago, is the youngest rock exposed; it forms the rimrock of the canyons westward nearly to Hoover Dam.

Major Powell noted this abrupt upfold in the rocks, as well as a second fold, the East Kaibab monocline, which the river crosses near the mouth of the Little Colorado River. This upfold brings ancient metamorphic rocks to the surface and forms the Grand Canyon proper. Thus, geologic forces created the uplands in two great stair steps through which the Colorado River has been forced to erode its canyons. Because the rock layers dip gently toward the east and the river flows downhill toward the west, rocks visible at river level become older as a traveler progresses downstream.

Major Powell believed the land was being uplifted at the same rate at which the river was down-cutting. We now know that this cannot be the case because the folded uplifts are much older than the river. The last folding of the monoclines took place about 65 million years ago, at a time when marine waters covered this region. For the past 30 million years or so, since the rivers have taken stable courses, down-cutting has exceeded uplift.

One recent interpretation has the original Colorado River flowing westward along its present course to Kanab Creek and then northwest to some unknown destination in the vicinity of the Uinkaret and Shivwits plateaus. It was later captured by the headward erosion of a westward-draining river system and assumed its present western course. Whether or not this scenario proves true, evidence from the Lake Mead area confirms that the river eroded the Colorado Plateau and reached its present-day western exit between 4 and 6 million years ago (Lucchitta 1990).

Floating down the Colorado from Lees Ferry, river travelers meet their first significant rapid in Marble Canyon at the mouth of Badger Creek (photo pair 2). Like nearly all rapids in the canyon, this one formed when flash floods caused by thunderstorms sent debris-laden mudflows containing boulders the size of trucks pouring down the usually dry tributary wash. The mud soon washed downstream, but the boulders remained, partially damming the river, constricting its flow, and thus increasing the local gradient. Eventually, if unhindered by man-made dams, floodwaters and strong currents can move the boulders around or even carry some of them away downstream, so that the rapids shift in configuration. New rapids are born, and older ones become more docile.

Nowadays there are no great surges to shift the rapids around. Since the floodgates of Glen Canyon Dam closed in 1963, the river has only once reached a flood stage of 100,000 cubic feet per second. That happened when exceptionally heavy snows in the Colorado and Wyoming Rockies melted rapidly in the spring of 1983, catching Lake Powell already nearly full and taking its controlling computer by surprise. The water could not be discharged fast enough and overflowed the top of the dam—which was hastily extended by adding flimsy walls of plywood sheets. Floodwaters nearly destroyed the dam as they raced over it and through its eroding flood gates for several weeks into Marble and Grand canyons below.

The flood of 1983 allowed hydrologists to see firsthand its effects on the river's rapids. Crystal Rapid, for example, at mile 98.1, had formed as recently as 1966 as a result of heavy rains on the North Rim. Because of its youth and the dam's lowering of flow rates since the rapid's formation, it had not yet become fully stabilized. (Rapids are generally considered to be stable and mature when the river's channel, at first partially or completely dammed by mudflows, has again been freed for about half its breadth above and below the obstruction.) In 1983, boulders that had created boat-swallowing holes in mid-rapid were shunted aside, and huge waves developed at the head of the rapid, flipping forty-five-foot river boats end over end. When the flooding subsided, Crystal Rapid had been shifted entirely to the left side of the river, leaving the right half a smooth, easy run, at least in low water.

The effect of the 1983 flood on Crystal Rapid dramatizes the importance of river torrents in stabilizing dangerous rapids. Considering the low maximum discharges released by Glen Canyon Dam, one can imagine that some future debris flow coursing out of a side canyon could shut off river travel completely because it would require centuries of unnaturally reduced flows to bring the new rapids to maturity.

At the foot of the treacherous Hance Rapid, the river enters upper Granite Gorge (photo pair 28), a dark, somber chasm that reaches thousand-foot depths beneath the Great Unconformity—the erosional plane separating upturned Precambrian sedimentary layers and older metamorphic rocks from younger, Paleozoic strata above. The age of the metamorphic rocks in Granite Gorge, 1.8 to 2 billion years old, was the time when the geological clock was last reset—that is, when all preexisting rocks and minerals in this complex recrystallized for the last time and the dating of all previous geological events became impossible.

These ancient rocks, generally referred to as the Vishnu Schist, were once deposited as sedimentary layers interrupted by lava flows. Then, in episodes of great heating and compression, the original rocks were nearly remelted and deformed into the contorted mass we see today. Finally they were shot through with granitic dikes and sills now known as the Zoroaster Granite (photo pair 29). More mountain-building forces tied the granitic dikes into bowknots and other preposterous shapes.

The Colorado crosses and exposes these metamorphic, crystalline basement rocks three times in Grand Canyon: first in upper Granite Gorge and again in middle and lower Granite Gorge downstream. In each case, the hard, generally homogeneous rocks firmly resist lateral erosion, so that the canyon narrows to a nearly vertically walled slash. The world closes in, and the traveler comes as close to the center of the earth as anyone can imagine.

THE COLORADO ROUNDS a gentle bend at mile 178 and is met by a mid-stream sentinel of black rock—Vulcan's Anvil (photo pair 41). This peculiar chunk of unexpected lava, apparently the neck of an ancient volcano, warns travelers of Lava Falls, the granddaddy of all Grand Canyon rapids. Vulcan's Anvil ushers in views of lava flows that cascaded down the canyon walls only about a million years ago after most of it had been carved by erosion. Seeing them, Powell wrote in 1875: "What a conflict of water and fire there must have been here! Just imagine a river of molten rock running down a river of melted snow. What a seething and boiling of waters; what clouds of steam rolled into the heavens!"

From Lava Falls, one can see hardened flows of lava that cascaded down the canyon walls from Vulcan's Throne, a beautifully preserved cinder cone on the canyon rim three thousand feet above. Other volcanic vents along the rim and within the canyon emitted molten rock from feeder dikes now visible in the canyon walls. More than 150 lava flows left deposits of basalt that can be seen downstream for eighty-five miles. These flows formed at least a dozen natural dams; their reservoirs extended upriver in some cases above present-day Glen Canyon Dam and into Utah (Hamblin 1990).

Although its name suggests that the rapid called Lava Falls was formed by remnants of lava, that is not the case. Instead—in the usual process—mudflows and rock debris from adjacent Prospect Canyon constricted the river's flow to form the most treacherous and feared obstacle to river travel in Grand Canyon. It was at Lava Falls that the 1923 survey party was stranded by floodwaters, as shown in photo pair 42. Few of the canyon's rocks, of course, sprang from the earth as spectacularly as its volcanic ones. Some varieties precipitated quietly out of mineral-bearing water into which still other rocks had dissolved. Just below Lava Falls, for example, lime-laden spring water gurgling from the Toroweap fault has produced extensive travertine deposits that encrust the south canyon walls.

Many rocks contain water; the amount depends on the volume of pore space in the rock. Sandstone, for instance, has more microscopic pore space between grains than one would think, and it usually is intricately interconnected to permit the water to move, if ever so slowly. Limestones are different: they usually have little natural pore space, but if they are attacked by rainwater, which is invariably somewhat acidic, the lime, or calcium carbonate, of the stone dissolves readily. As it does, underground drainage passageways and even caves may develop—such as those found at Vasey's Paradise in Marble Canyon (photo pair 7).

Where does the lime go that is dissolved from the limestone? It commonly forms deposits like those seen inside Carlsbad and Mammoth caves, but much of it emerges from the rock via springs and seeps, still in solution in the water. Evaporation then causes it to precipitate in the form of travertine deposits.

Travertine, a white, brittle coating seen on the canyon walls and in tributary streambeds, is common in side canyons that are fed by springs and seeps, especially those emitting from the Redwall Limestone. Unless contaminated by floodwaters from upstream, these spring-fed tributary waters appear to be sky blue, a result of light dispersion and reflection of sky colors by the tiny white crystalline particles of calcite.

One of the springs that feeds the Little Colorado River gushes calcite-laden, sky-blue water in a nearly closed travertine dome that Hopi Indians believe to be the *sipapu,* the opening through which ancestral Holy People emerged from the underground world. Other noticeable travertine deposits decorate Havasu Creek, Pumpkin Spring, and Travertine Falls in Grand Canyon. Deposition of travertine in Havasu Creek is so relentless that it forms tiny dams of cemented grass and twigs, frozen in place even as they remain green with life.

Back upstream in lower Marble Canyon, seepage from the Bright Angel Shale has cemented the talus slopes along the river into surface rock that is just now being gullied by modern erosion. Such travertine-cemented talus deposits can be seen especially well near the mouth of the Little Colorado (photo pairs 19 and 20).

If geologic change has been negligible since 1923, changes in vegetation along the Colorado's shoreline have been remarkable. Before Glen Canyon Dam, annual floods raked Marble and Grand canyons, mostly during May and June. The canyon walls up to the high-water line were stripped nearly bare of vegetation. Above the high-water line, the most common plants were cottonwood, mesquite, arrowweed, and willow; they have been little affected by the construction of the dam. At lower elevations, however, now that floods are rare, vegetation has a chance to take root and become well enough established to withstand the few floods that do occur. A whole new assemblage of plant life flourishes at river's edge.

The most noticeable of the recent invaders is tamarisk (*Tamarix chinensis* Lour), or salt cedar. Tamarisk is sometimes valued as a decorative shrub, but to river travelers along the Colorado, it is a scourge.

Tamarisks are phreatophytes, as are willows and cottonwoods. (Sandbar willow, another phreatophyte, often competes with tamarisk for dominance along sandy shorelines.) They absorb great amounts of water through extensive root systems, some as much as a hundred feet deep, and then pump the moisture freely into the air through their wispy leaves—thus lowering the water table and exaggerating the effects of normal evaporation. Their root systems trap and bind sand flats, and by fixing these sediments narrow the course of the river and increase water velocities in otherwise quiet stretches. The same effect stabilizes sandbars and fixes low-water meanders into semipermanent patterns (photo pair 17).

Native to lands bordering the Mediterranean Sea, the willowy tamarisk, with its pretty, tiny, purplish flowers, came to North America as an immigrant. Perhaps the most plausible story of its arrival is that Spanish priests imported tamarisk to decorate the gardens of their missions in Mexico. Indeed, the plant's generic name is derived from that of a river in Spain. Exactly when the new species appeared in the western world is anybody's guess—perhaps during the eighteenth or nineteenth century.

Salt cedar was first recognized in the American Southwest in about 1900, along the Salt River in southern Arizona (Turner and Karpiscak 1980). It was spotted along the Virgin River in Nevada by 1925 and along the Colorado River near the mouth of the San Juan in the mid-1930s. Its rate of invasion probably picked up between 1935 and 1955 (Turner and Karpiscak 1980:14).

Photos taken by Birdseye's 1923 survey expedition show very few tamarisks along the Colorado in Grand Canyon. By 1938, Elzada Clover and Lois Jotter, botanists and the first two women to make a successful run through Grand Canyon, noted isolated tamarisks along much of the length of the canyon (Clover and Jotter 1944).

Tamarisks can be prolific, dispersing almost microscopic seeds that are carried throughout the Colorado River basin on the prevailing winds. The nearly ubiquitous seeds require moist, stable conditions to establish growth, especially during the summer months. Before dam construction, floods prevented such stable sediments from accumulating, so the nearly continuous riverbank groves we see today must have developed after river levels were lowered by regulated dam releases (Turner and Karpiscak 1980).

Except for providing homes for insects and small birds, the prolific growth of tamarisk is of dubious value. Tamarisks tend to outcompete other plant species, and, from the river runner's perspective, they destroy otherwise pleasant campsites for recreational users.

Scientists are studying ways to control the spread of tamarisks along the river banks—witness the number of trees tagged for monitoring. Controlled burning of tamarisk groves in Canyonlands National Park was found to be ineffectual; the trees returned with renewed vigor the following season. No insect or bacterial species have been identified that will destroy or slow the tamarisk's spread, nor has an otherwise environmentally safe chemical been discovered for its destruction. Trampling by recreational users of the canyons is only temporarily effective. For the foreseeable future, tamarisks will have to be tolerated.

Other exotic plants have invaded the canyon floor in historic times, though none has become so profuse or bothersome as the tamarisk. Bermuda grass, native to Eurasia, was reported in the lower canyon by botanists Clover and Jotter (1944), and probably now occurs throughout most of the canyon (Turner and Karpiscak 1980). Russian olive was not found in the canyons until 1973, although the thorny, distinctive, olive-green trees are known to spread rapidly and may become a nuisance in future years.

Camelthorn, accidentally imported from Asia in about 1930, grows in scattered stands below the mouth of the Little Colorado above the dam-controlled high-water line (Turner and Karpiscak 1980:15).

Plants that have become widespread since dam construction between the old and new high-water levels are longleaf brickellia, desert broom, and waterweed. Hackberry and rosebud trees grow chiefly above the pre-dam high-water line, especially in north-facing and seep-moistened alcoves, but they have spread to lower levels at such locations as Vasey's Paradise and Lava Falls. Cattails and reeds, once rare, have become common in quiet, marshy locations since dam construction.

Over the course of several million years, geologic processes produced the canyon as it was viewed by members of the Powell and Birdseye expeditions. Since then, human usage has greatly accelerated the rates of change. Our concerns for protection of the canyons and the fragile desert environment from human depredation are appropriate and necessary to provide a nearly natural showcase for future generations to enjoy.

The Canyon Revisited 1923/1991

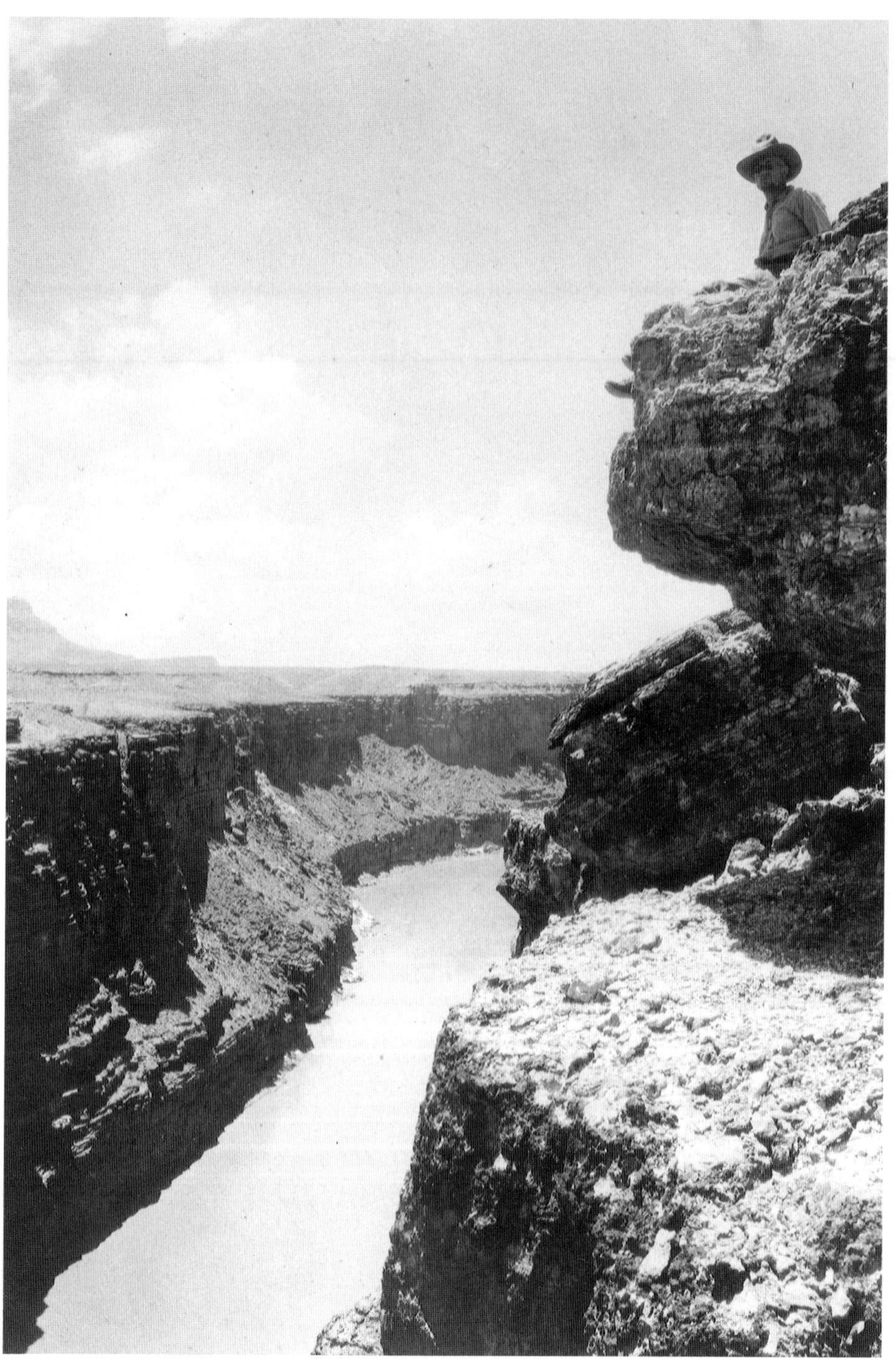

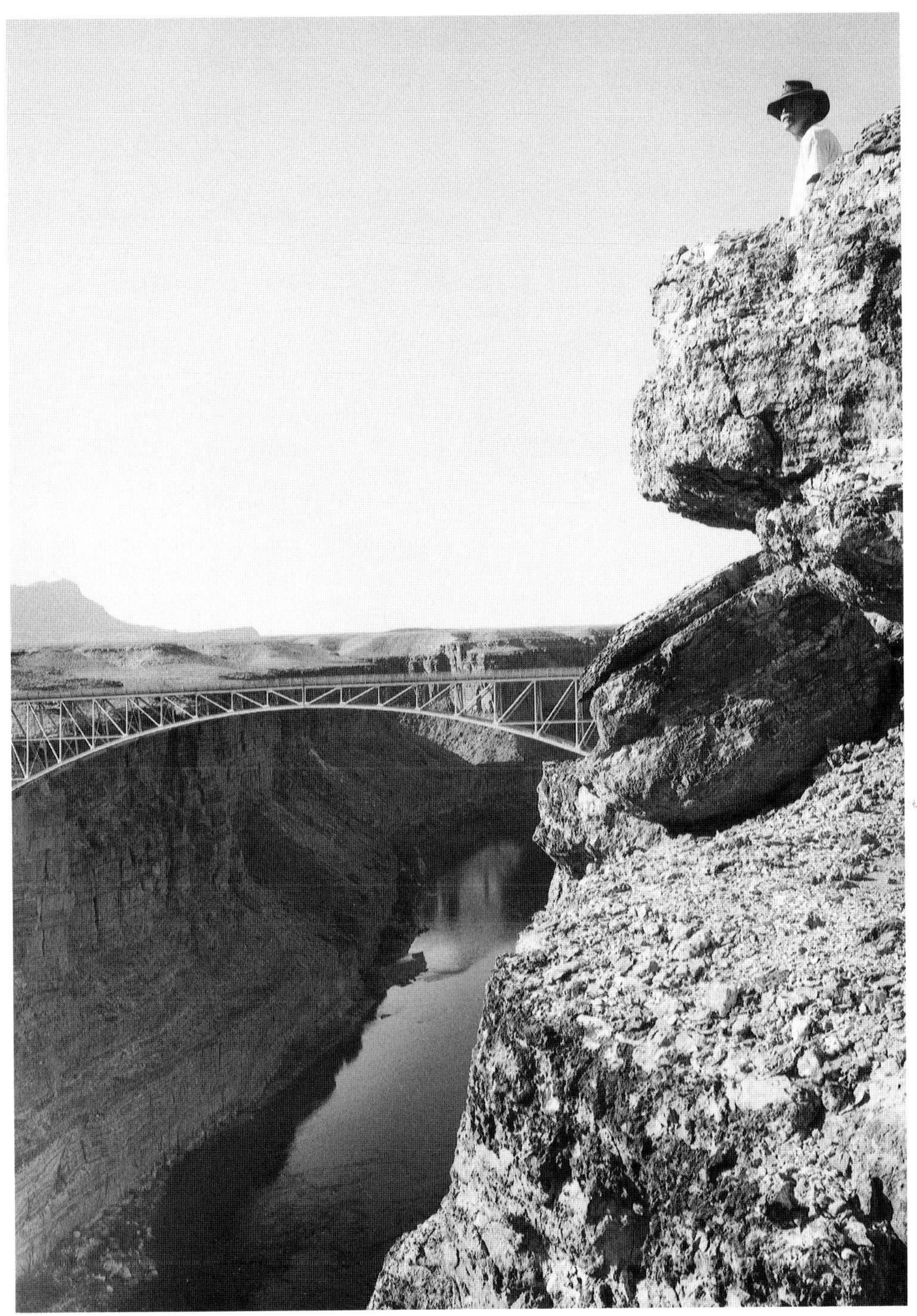

1. Navajo Bridge, mile 4.1, September 24, 1991, 9:00 A.M.

1. Navajo Bridge, mile 4.1, September 24, 1991, 9:00 A.M.

When Claude Birdseye's survey party reached the head of Marble Canyon in 1923, no bridges spanned the Colorado River south of Moab, Utah. Other than the dangerous Crossing of the Fathers, a ford pioneered by Spanish explorers upstream from the present site of Glen Canyon Dam, the only means of crossing were ferries. The nearest one was Lees Ferry, about four miles above the present-day site of Navajo Bridge.

The most obvious change between these two photos is, of course, Navajo Bridge, built in 1928. It sits 467 feet above the Colorado and spans 834 feet. In 1923, the survey party knew where the bridge was to be built, and R. C. Moore drew a cross section of the canyon walls there. At this writing, a new bridge for vehicular traffic is planned a short distance downstream. The current Navajo Bridge will be retained as a pedestrian crossing and lookout point for viewing the river.

This pair of photos was taken one-quarter mile north of Navajo Bridge, looking south-southeast down Marble Canyon. The Echo Cliffs are in the background to the left. This vantage point lies about twenty miles downstream from today's Glen Canyon Dam.

In the older photo, R. C. Moore is perched on a ledge of Kaibab Limestone. Lewis Freeman took the photo about midday, before the party set out down the Colorado. While the rest of the crew prepared the boats, Moore and Lewis Freeman studied the landscape above Marble Canyon. An entry in Moore's field notebook for July 19, 1923, reports: "Top of Marble Canyon observed. Ss [sandstone] cliffs north of river very fine. Chinle covered by talus and some slump . . . All brightly colored."

Tom McClain is in the recent photo, although he was not keen about hanging his feet over the edge. The new photograph shows negligible changes in the cliff on the righthand side of the picture. In the right foreground, several pieces of talus rock are now missing, while others still lie exactly where they were sixty-eight years earlier. The steep cliff face on the left side of the photo is roughly the same, with many of the talus slopes and large boulders still in place. River level in the 1923 photo appears to be somewhat higher than in the recent shot. The sandbars on the left side of the river are more pronounced in the older photo, probably because of beach erosion and the river's inability to supply new sediment after the construction of Glen Canyon Dam.

The upper cliff-forming rock unit is the Kaibab Limestone, the intermediate slopes are on the Toroweap Formation, and the lower cliff is formed on the Coconino Sandstone.

Although this set is the first of the photographic pairs reproduced in this book, our photo was actually the last one we took in 1991. After completing the trip downriver, we returned to Lees Ferry to spend one more night. We located the vantage point late that evening, then went back to take the repeat photo the next morning when the sunlight more nearly approximated that of the original photograph.

2. Badger Creek Rapid, mile 7.8, September 8, 1991, noon.

Badger Creek Rapid is the first rapid encountered in Marble Canyon. It formed sometime before 1869, when it dealt danger to the first river exploration party led by John Wesley Powell. Mudflows from the tributary canyon of Badger Creek (so named because the Mormon scout Jacob Hamblin once shot a badger there) had restricted the river's flow and left large boulders along the far bank and in midstream. By 1923 the rapid had not been reduced in authority and still raged a warning to the surveying boaters. Note the *Boulder* running the rapid stern-first in the older photo.

Perhaps obvious only to a Grand Canyon boatman, the large waves and related holes in the rapid in the 1923 photo were created by large underwater boulders. There is no hint of these boulders in the 1991 photo, despite the water's being lower. Apparently they have been removed or shifted to less menacing locations. Clearly the rapid was much more difficult to navigate in 1923 than it is today, justifying the earlier crew's great respect for the river.

The original photo was taken on August 2, 1923, looking to the west-southwest up Badger Canyon from a boulder at the edge of the river's east side (it is still in nearly the same position). On August 1, Moore stopped to draw a cross section here and identified the rocks on the opposite wall as (from the bottom) the Hermit Shale (the dark shale visible where the canyon slopes come together in a V), the Coconino Sandstone, and the Kaibab Limestone; the Toroweap Formation of the lower cliffs had not yet been distinguished and named in 1923.

Perhaps the most obvious difference between the two photos is the increase in vegetation, primarily tamarisk, at the mouth of Badger Creek. Indeed, the vegetation is so thick that it obscures many of the boulders at the mouth of the creek and makes comparison difficult.

The high canyon walls appear unchanged in the two photos. Several large boulders with their flat sides to the camera are in the same position high up on the talus slope toward the left of the photos. Some very large boulders on the nearer talus slope also appear unmoved. A few new, small boulders, however, appear on the near talus slope in the recent photo, having fallen from the cliffs above since 1923. These new boulders are lighter in color because they have not had time to weather and oxidize to the same color as the surrounding rocks.

The sand beach at the right in the 1923 photo is now missing, and the one to the left is considerably reduced, a characteristic trait of beaches in the upper canyons. When Glen Canyon Dam was completed in 1963, the sediment supply to the river was essentially cut off, with the result that beaches and sandbars in the upper canyons now erode and some of their sediments are redeposited in the lower canyon.

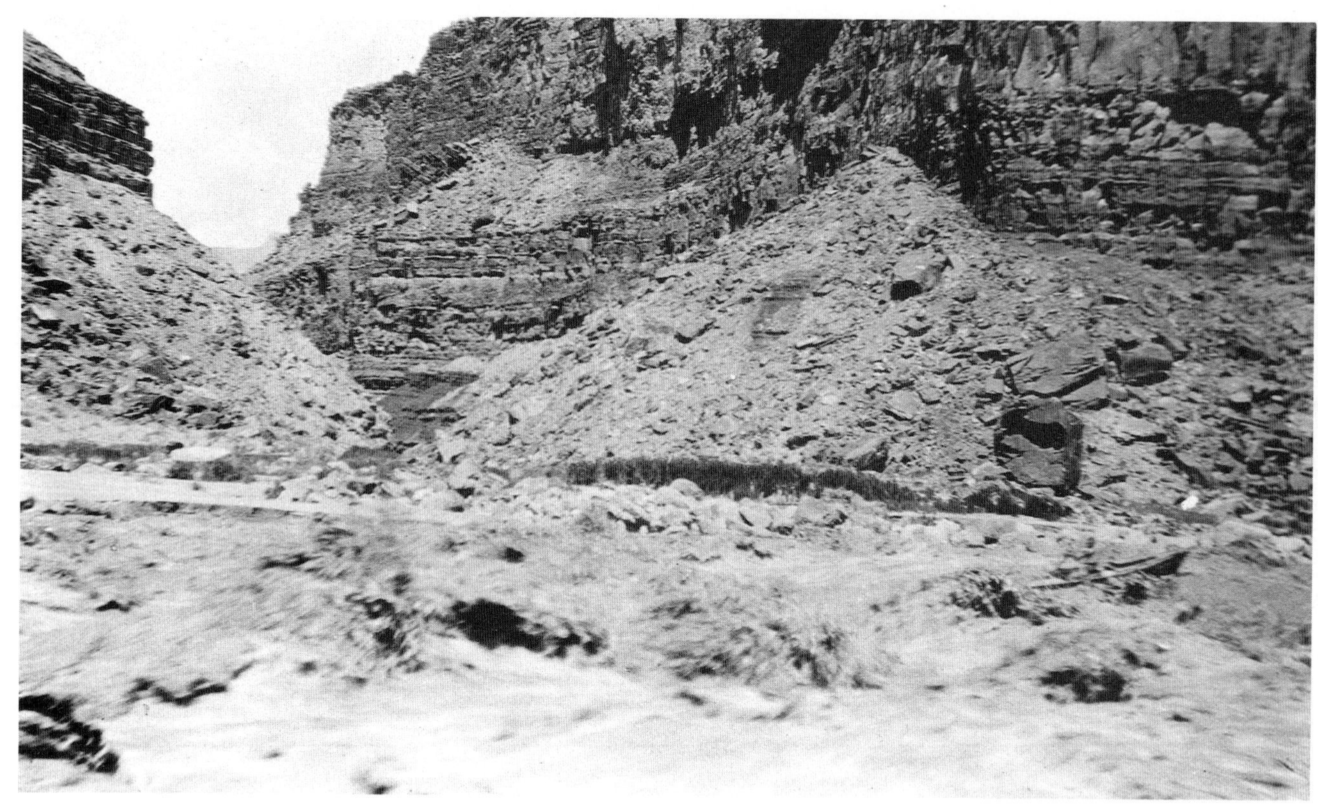

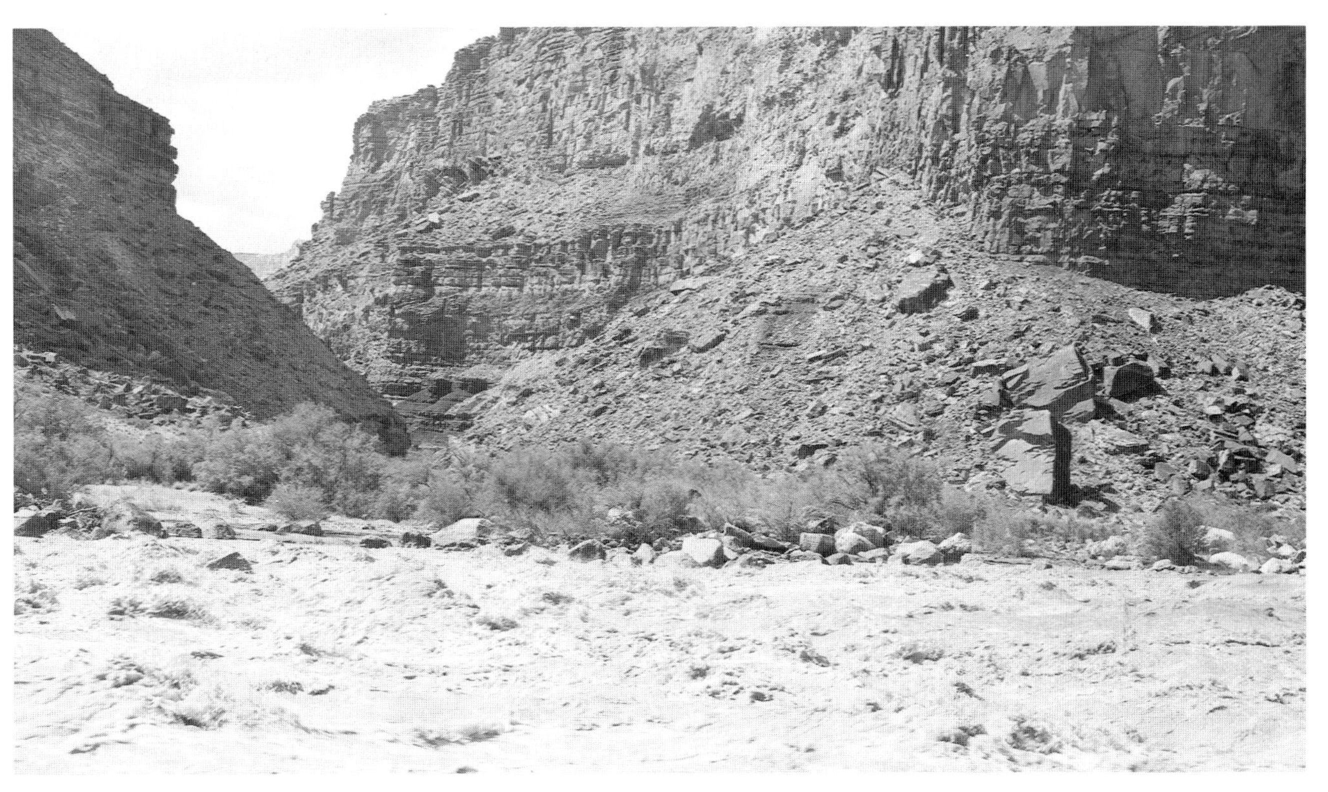

2. Badger Creek Rapid, mile 7.8,
September 8, 1991, noon.

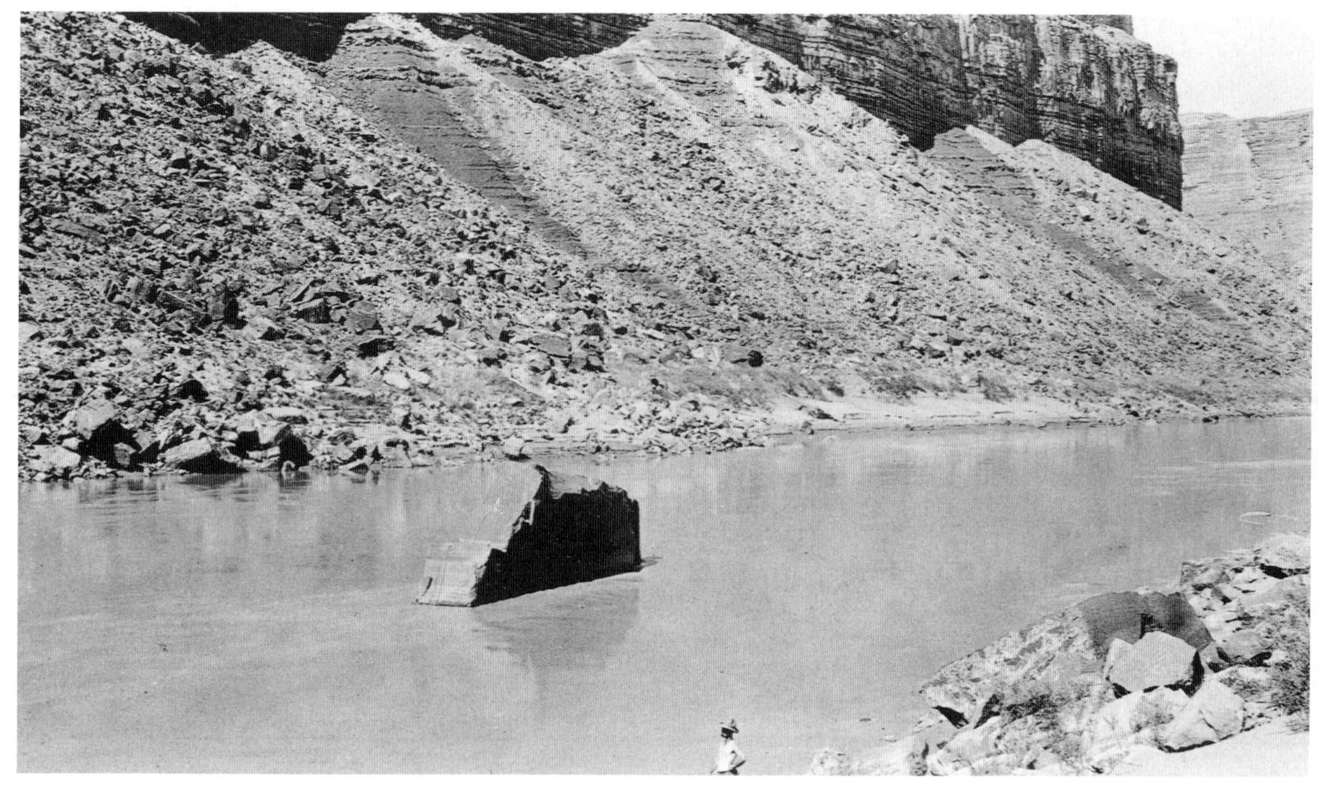

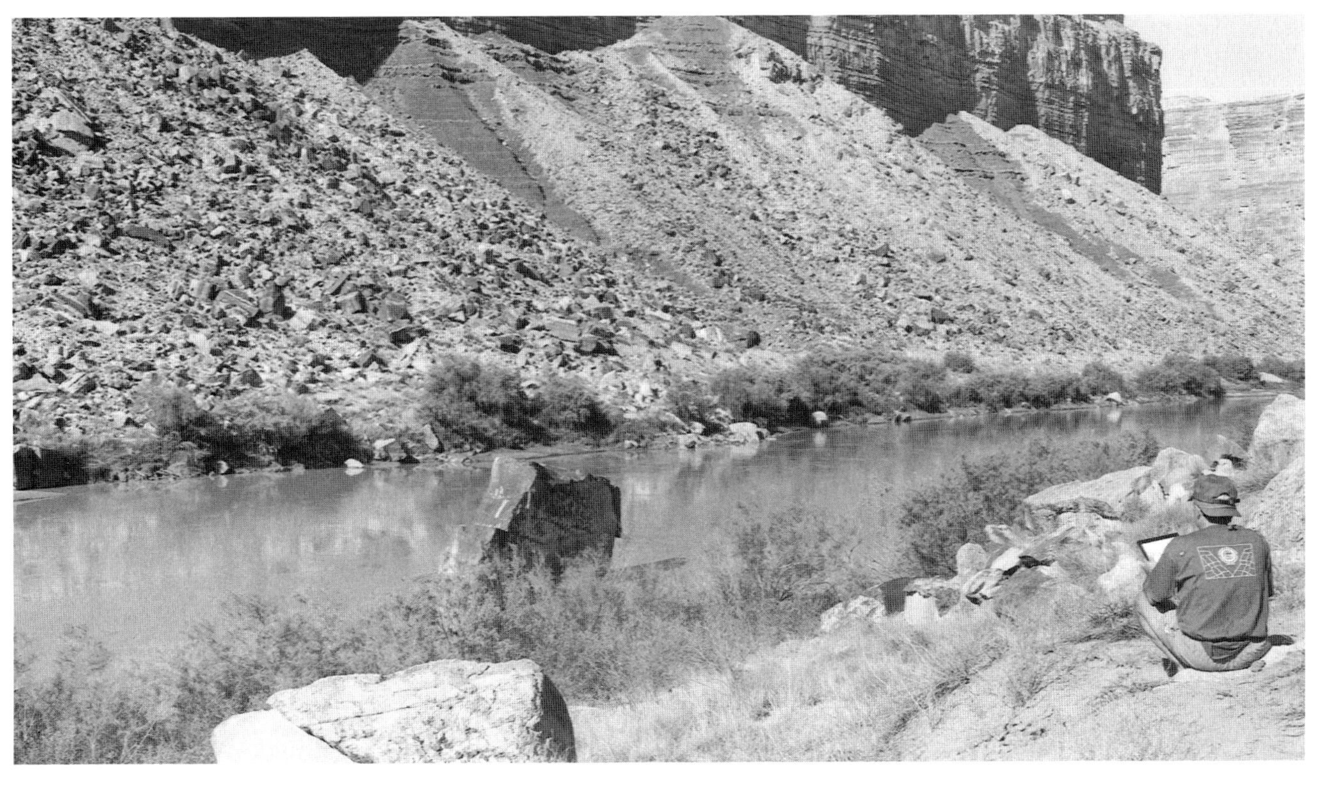

3. 10-Mile Rock, mile 10.0,
September 8, 1991, 3:00 P.M.

3. 10-Mile Rock, mile 10.0, September 8, 1991, 3:00 P.M.

Near the ten-mile point on the river rises a slab of limestone that tumbled off the cliff face and landed on edge in the middle of the Colorado. About fifteen feet of the rock protrudes above water level, although the entire slab is probably about forty feet long.

The 1923 photo of 10-Mile Rock was taken in mid-afternoon on August 2, 1923. Moore drew a cross section at this spot, calling it in his notebook a "sketch of canyon wall just below big boulder block midstream." Artist Louis Copt is in the newer photo.

Again, the water level was somewhat lower when the modern photo was taken, and the far cliff faces and talus slopes appear largely unchanged. When unaffected by river flow or human activities, talus slopes tend to stay put over the course of a few decades.

Closer to the river, however, one can see considerable change. In the new photograph, the zone of regular river fluctuation is now overgrown with tamarisks. Higher river levels are marked by a strip of light-colored rocks a few feet above the plants. Below that line and on the near bank, many boulders have been shifted.

It was impossible to line up the foreground precisely in the new photograph because much of the sand beach at 10-Mile Rock has washed away downriver (dam-related erosion again), and dune sand along the river's edge has shifted. The original vantage point lay about ten feet from the current edge of the sand beach. Moreover, the dense tamarisks along the edge of the river barricaded us from the exact location.

4. Canyon wall near Soap Creek, mile 11.1, September 9, 1991, 8:30 A.M.

Soap Creek enters the Colorado from the west. The creek and its rapid, visible in the lower left of the photo, got their names because it was here that Jacob Hamblin, camping overnight, boiled in alkali water the badger he had shot upstream, finding the next morning that the fat had turned to soap. The lower slopes of the canyon at Soap Creek consist of the Hermit Shale, protected from above by more resistant layers of Coconino Sandstone, the Toroweap Formation, and the caprock, Kaibab Limestone.

The earlier photo in this pair was taken in the late afternoon on August 2, 1923, from a sandy bench just above, or north of, the mouth of Soap Creek, looking south at the canyon wall. The differences between the two photos are not dramatic. Because the rocks of the canyon walls are so enduring, photographs taken from relatively high vantage points (farther away from the river) tend to reveal fewer changes than those taken closer to the river. In these views, most of the change takes the form of increased vegetation on the beach in the foreground, where tamarisks obscure many of the large boulders and make it difficult to discern that the boulders have, in fact, been rearranged by high water.

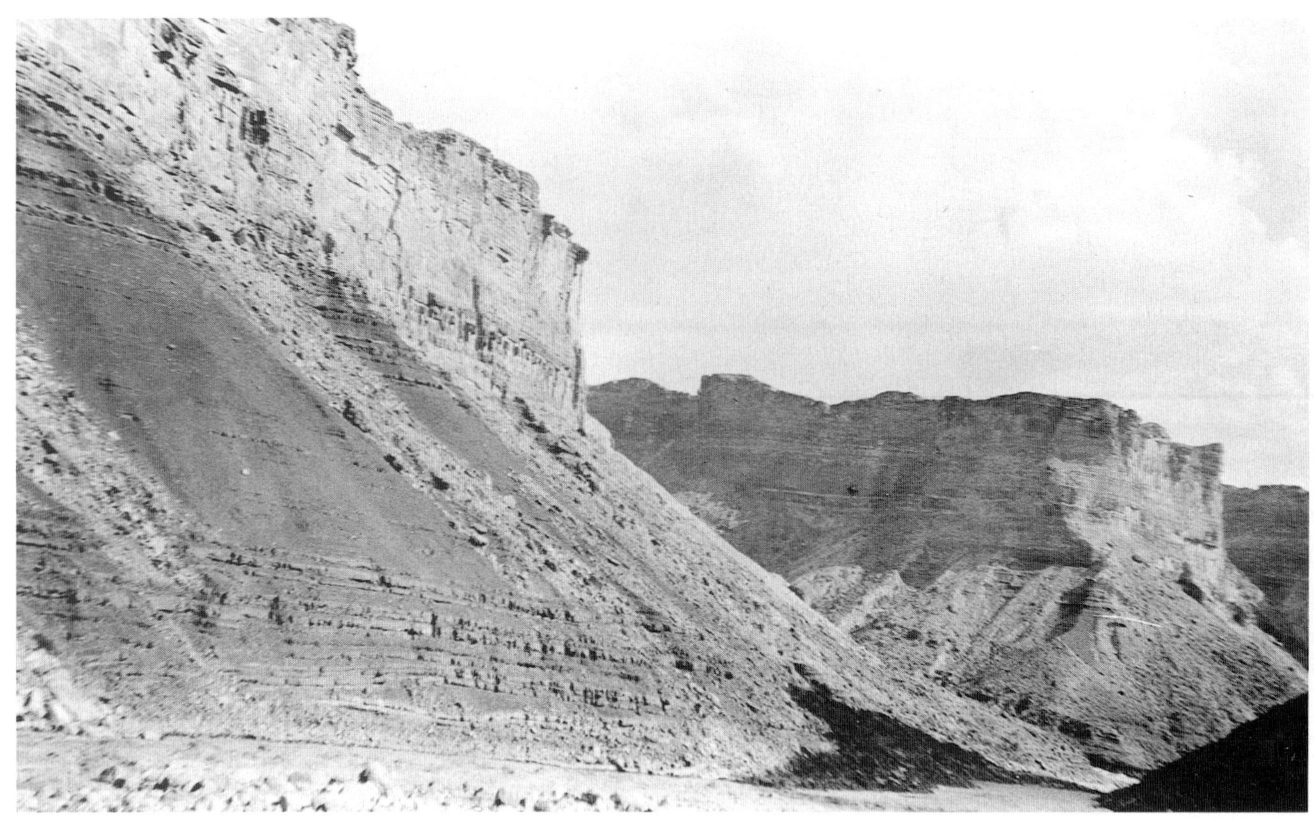

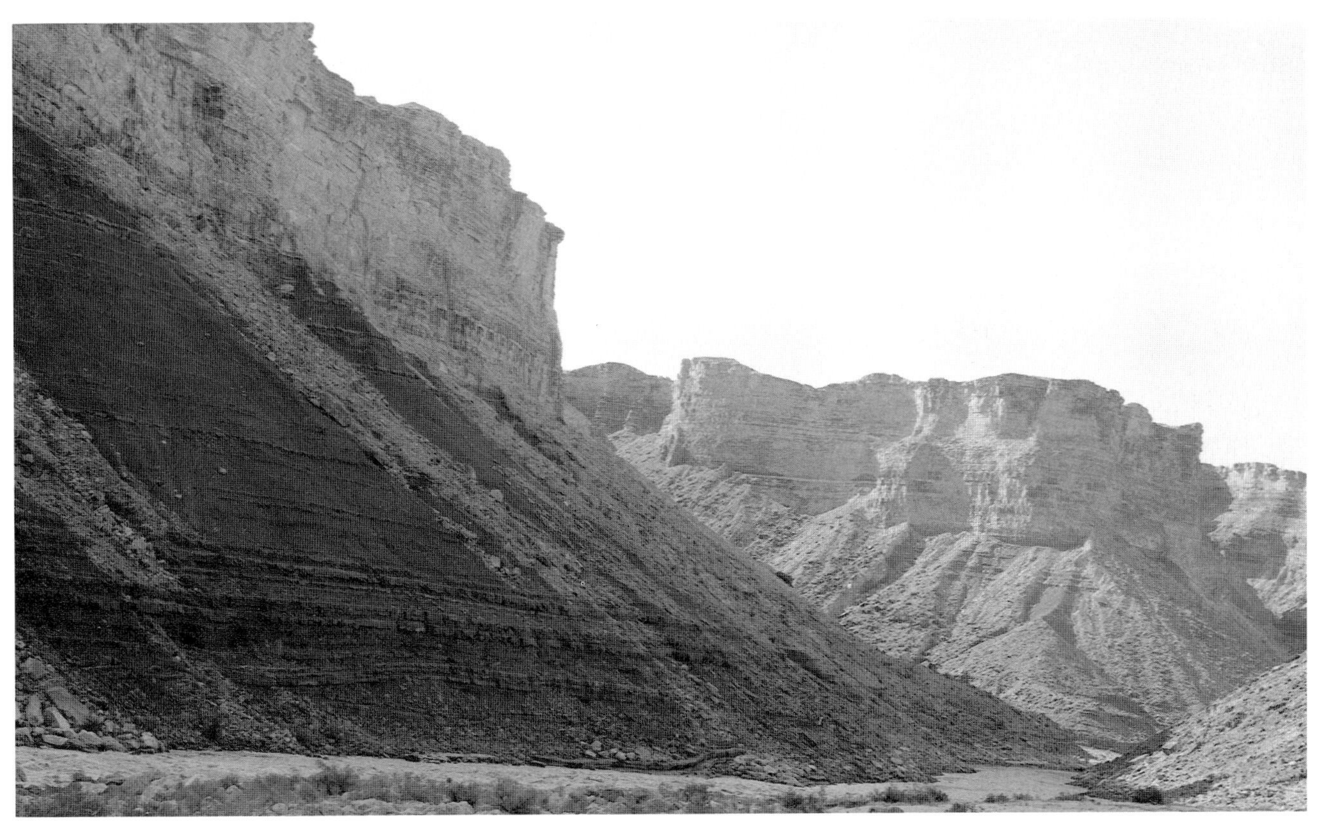

4. Canyon wall near Soap Creek,
mile 11.1, September 9, 1991,
8:30 A.M.

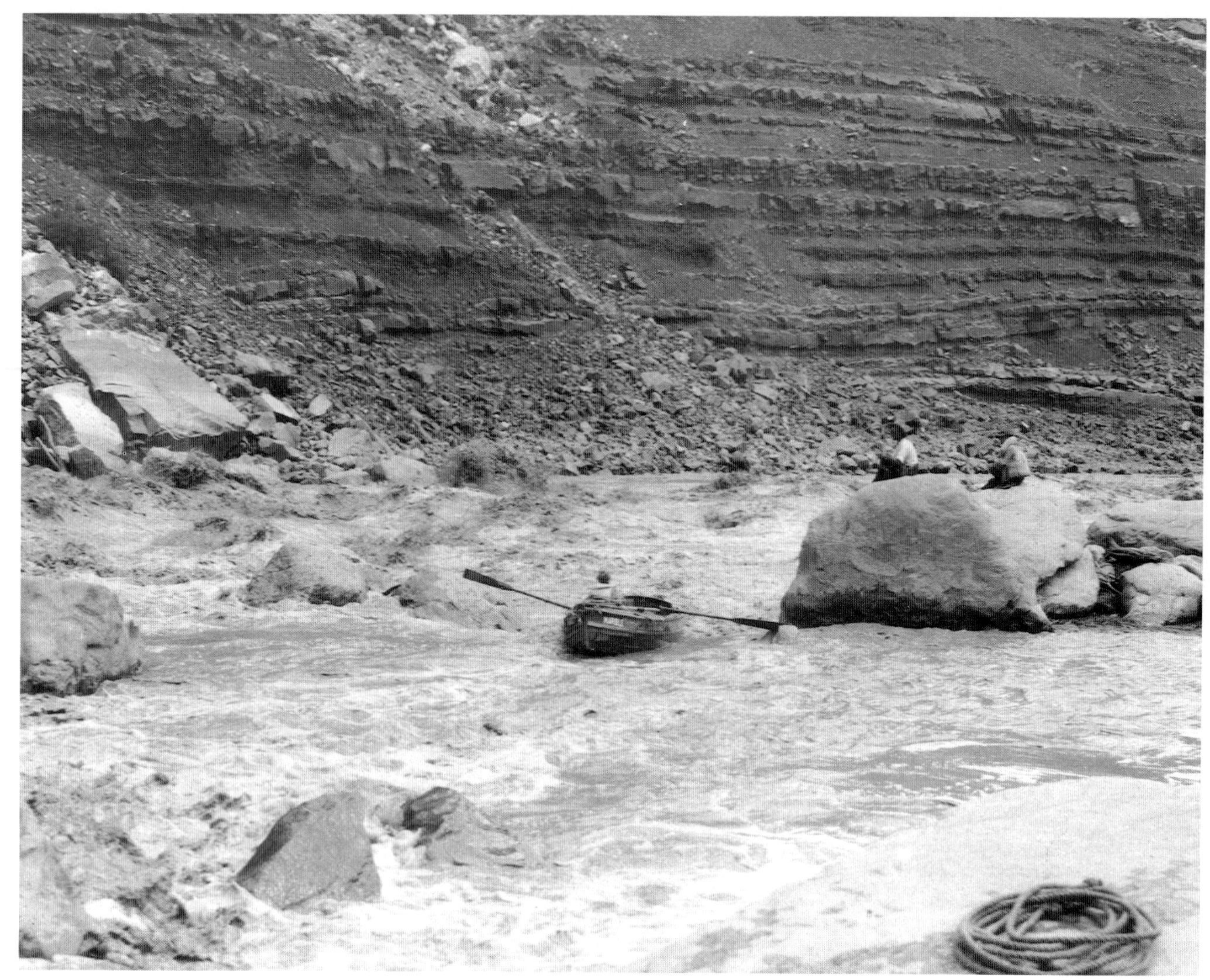

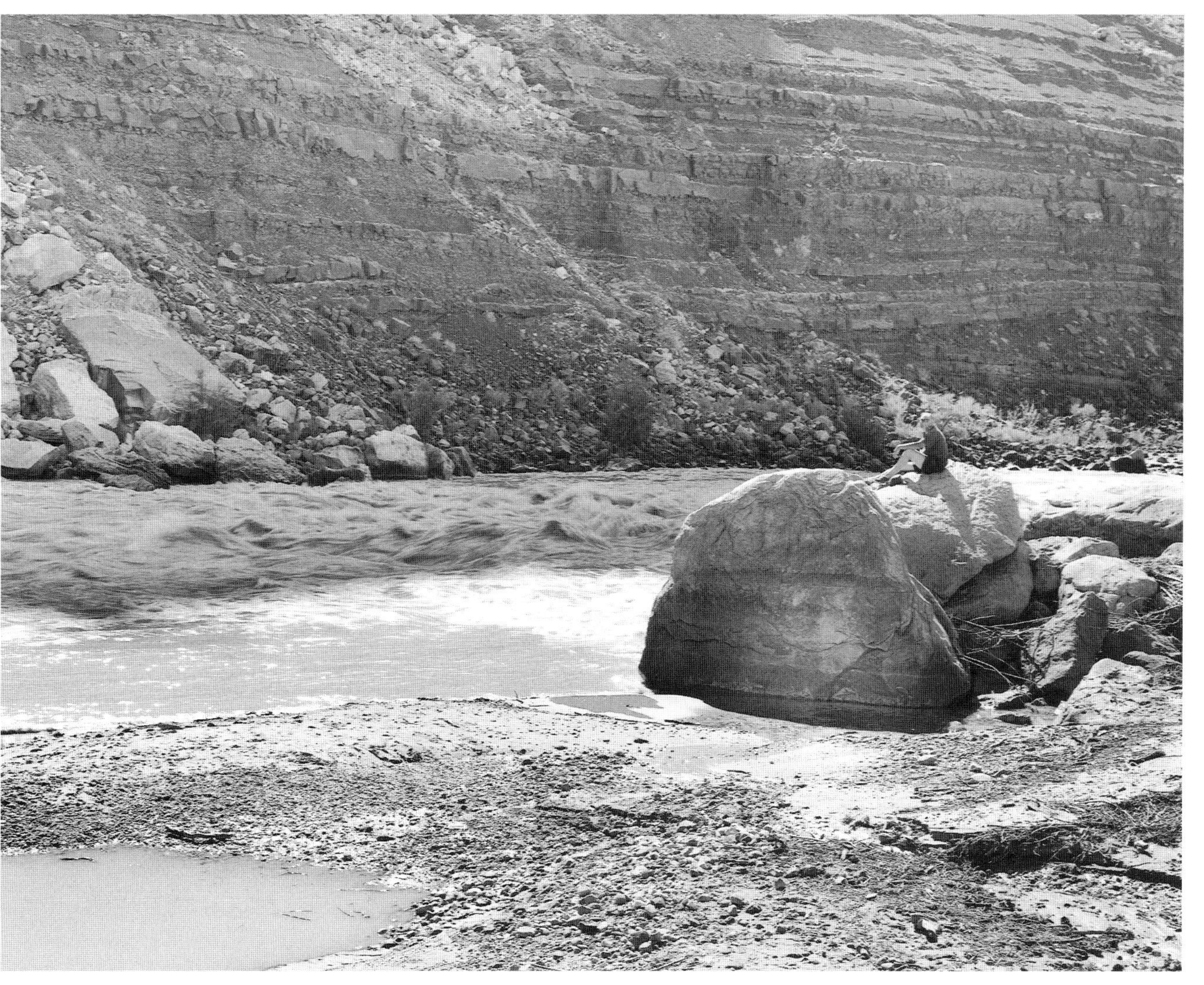

5. Soap Creek Rapid, mile 11.2,
September 9, 1991, 10:00 A.M.

5. Soap Creek Rapid, mile 11.2, September 9, 1991, 10:00 A.M.

Ellsworth Kolb, whose brother, Emery, was the lead boatman on the 1923 trip, described his ride through Soap Creek Rapid during their 1911–12 trip down the river. After being thrown from his boat into the cold Colorado, he wrote, "Somehow I had lost all desire to successfully navigate Soap Creek Rapid." Remembering that incident, Emery Kolb decided to portage the head of the rapid in 1923, and it was not successfully run until Clyde Eddy made it through in 1927. Soap Creek Rapid still demands healthy respect from modern river runners, but this photographic pair makes it obvious that the rapid is now a piece of cake compared with earlier times.

In the 1923 photo, taken from the west bank of the river at about midday on August 3, the boat *Marble* enters the lower rapid after being partially portaged around obstacles at the head of the watery turmoil. Two party members watch from safety; on the far canyon wall above their heads is a change in the strata of the Hermit Shale that Moore called an "irregularity."

The rapid had been created by boulders washed into the river from the tributary canyon of Soap Creek, directly behind the camera station. Despite the great size of the visible boulders, several of them had been carried away by pre-dam floodwaters by the time we took our photograph. It is easy to understand the hesitance of earlier boatmen to challenge the hazards then in place.

On the talus slope in about the center of the photos, a small boulder that was perched precariously in the 1923 shot has fallen prey to gravity by 1991. Vegetation is beginning to establish itself near the edge of the river, although strong currents and steep banks along this narrow stretch of the rapid have denied the ubiquitous tamarisk its foothold.

The recent shot was taken from slightly nearer the river than the original, owing to readjustment of boulders at the camera station. High-water marks are visible on the rocks where Dan Merriam is perched.

6. Soap Creek Rapid, mile 11.2, September 9, 1991, 1:00 P.M.

Birdseye's survey party camped at the mouth of Soap Creek on August 2 and 3, 1923. They had portaged around much of Soap Creek Rapid, then put in at this location to run the less dangerous final stretch. Their photo was taken about midday on August 3, from the south bank of the creek looking north-northeast.

In camp on the evening of August 2, the group set up their radio for the first time and received news of President Warren G. Harding's death. Two miles below here, at mile 13, expedition members studied closely the geology of two proposed dam sites.

Our 1991 photo captures only the tip of the large boulder in the right foreground of the older photo; we could not relocate the exact vantage point of 1923 because even huge boulders at this location had shifted. Although boulders to the right of the standing figure in the old photo remain in nearly identical positions, large rocks to his left have been replaced by smaller ones. It would take water velocities several magnitudes greater than those released from Glen Canyon Dam today to move boulders of this size. The boat *Grand* is in the older photo, in which the water level is considerably higher and vegetation, mostly tamarisk, is not yet established.

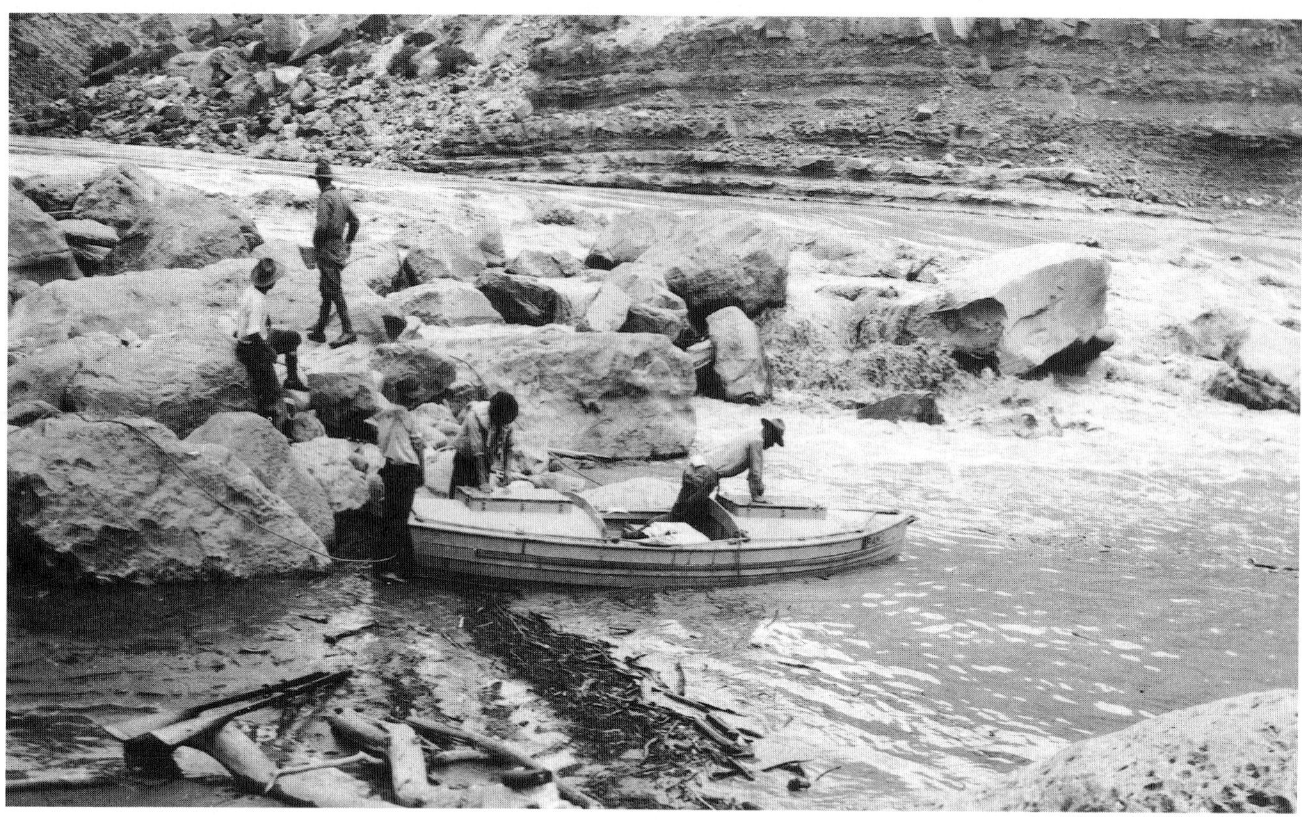

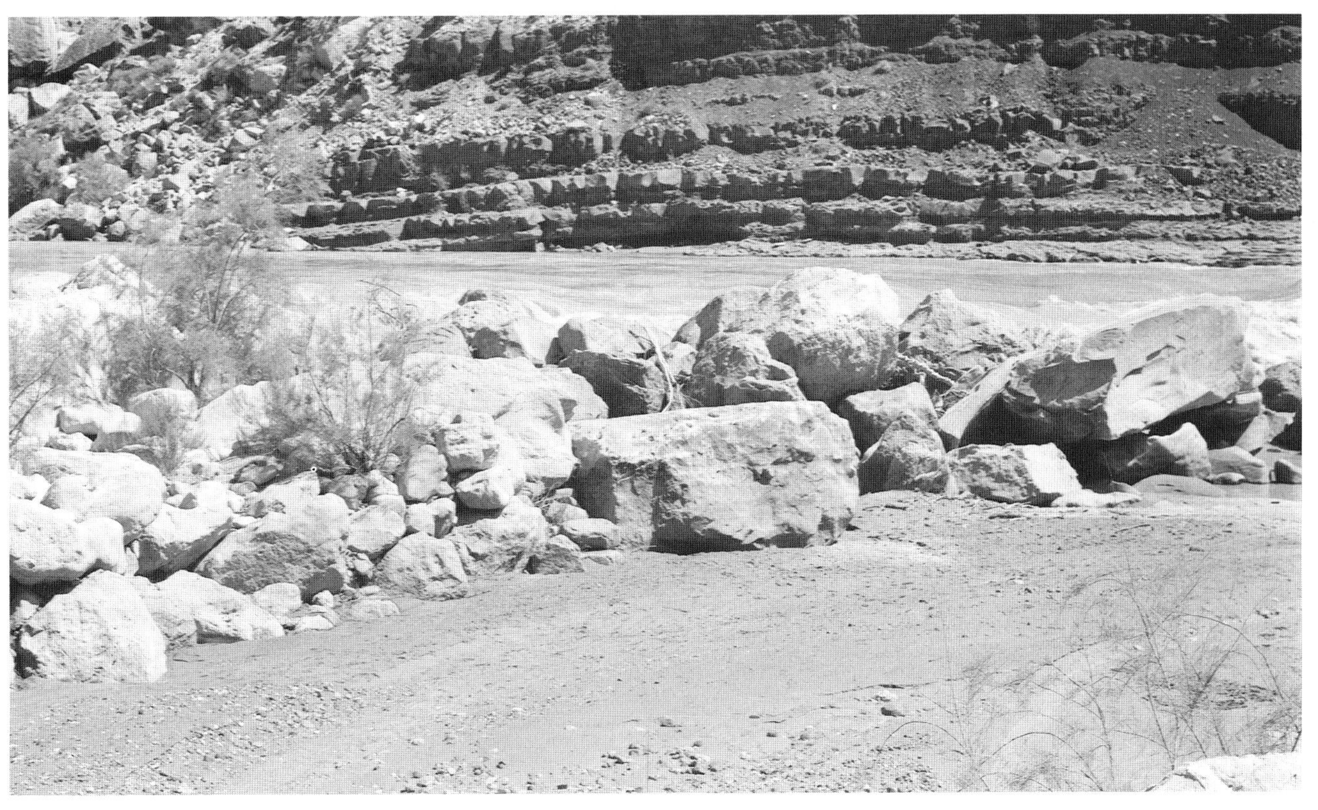

6. Soap Creek Rapid, mile 11.2,
September 9, 1991, 1:00 P.M.

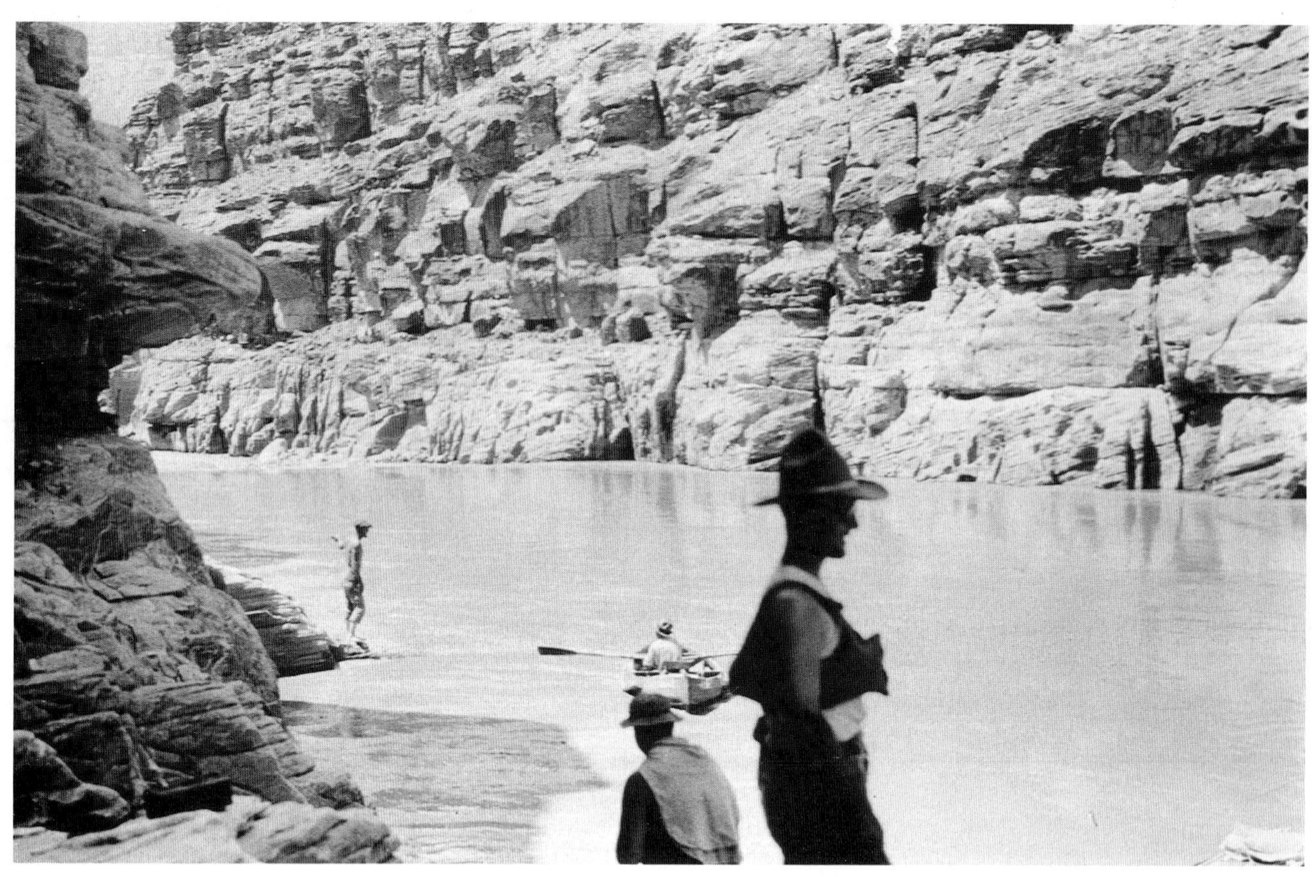

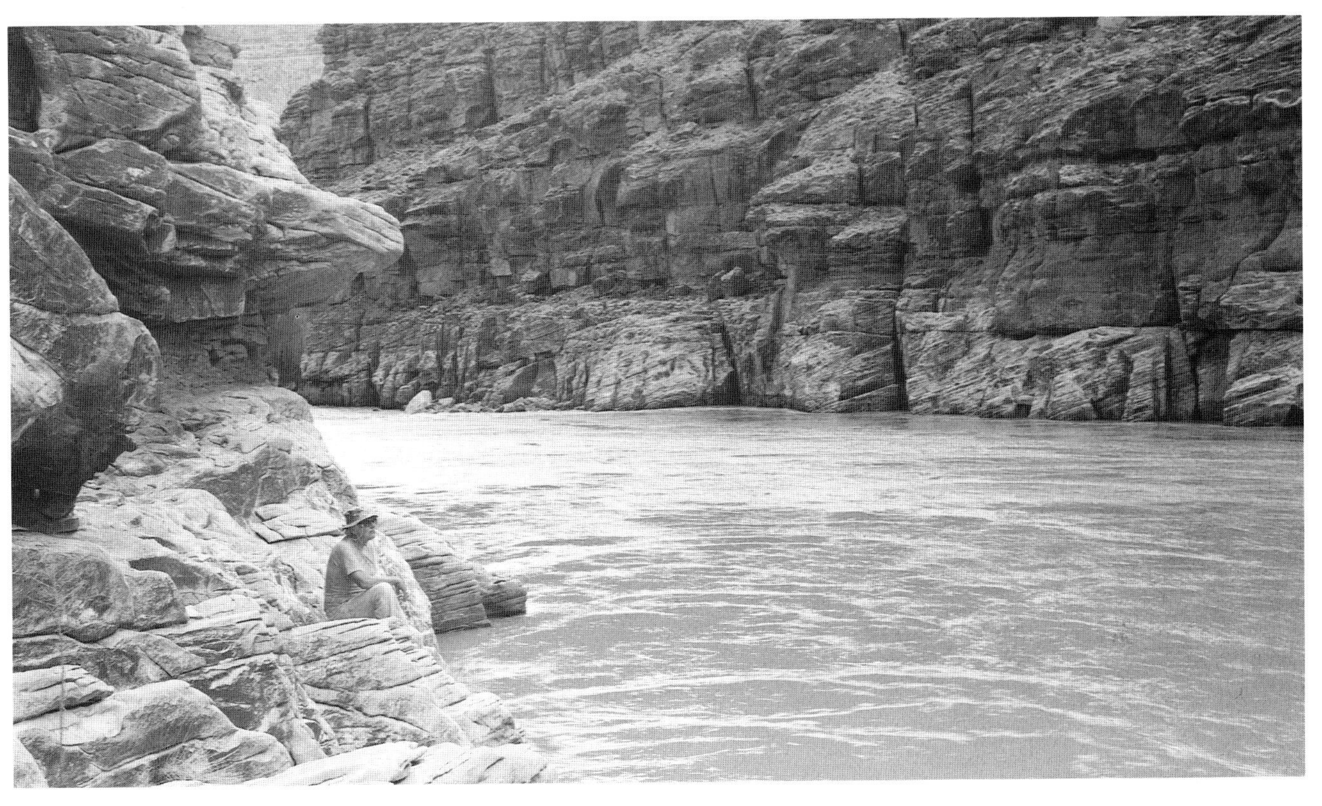

7. Near Tanner Wash
(Sheerwall Rapid), mile 14.4,
September 9, 1991, 3:30 P.M.

7. Near Tanner Wash (Sheerwall Rapid), mile 14.4, September 9, 1991, 3:30 P.M.

The 1923 party stopped here briefly about mid-morning on August 4 to select the next surveyor's turning point for the river transect and took this photo from the west bank, about ten feet above the river. The water level is slightly lower in our 1991 photo, but otherwise the two scenes are virtually identical.

Both photos show clearly the crossbedding in the layer of rock immediately above the water level—the Permian-age Esplanade Sandstone of the Supai Group. This well-hardened rock has been virtually immune to erosion during the last seven decades. These photos demonstrate how unimaginably slowly nature works its course on erosion-resistant rocks. In narrow stretches of the canyon like this, where the river runs wall-to-wall, human visitors rarely stop to disturb the pristine landscape, and there is no sand or soil to support even the hardy tamarisk.

Raymond Moore stands in the foreground of the 1923 photo, wearing a cowboy hat and life vest; Dan Merriam, once a student of Moore's, sits by the river in the 1991 version.

8. House Rock, mile 18.6, September 9, 1991, 6:00 P.M.

House Rock is the large block of Esplanade Sandstone that sits in the middle of the river below House Rock Rapid in these two photos. It has been a river-runner's landmark since the time of Powell's first voyage. The earlier photo in this pair was taken on August 5, 1923, on the east side of the river, looking back upstream to the north-northwest from a long, flat bench of sandstone of the Wescogame Formation (Supai Group). Moore drew a detailed sketch looking down Marble Canyon from high above House Rock Rapid.

Although changes on the far shore are minimal, on the near bank and talus slope they are substantial. Many of the rocks in the foreground of the 1923 photo have been washed away or rearranged, and new ones have fallen from the cliffs above. Farther upstream, several boulders on the talus slope above House Rock have rotated. A large driftwood log in the earlier picture has since washed away. Although the water level is lower in the newer photo, the shoreline where the boat is moored in the older shot appears to have receded somewhat by 1991, and the talus slope has built outward with more recent rockfall. Early morning shadows in the original photo obscure boulders that are visible in the newer one.

A small amount of vegetation, mostly tamarisk, now grows in the fractures in the lower ledge and between boulders on the talus slope. Their growth has been stunted, probably because they lie within the zone of daily fluctuating river levels caused by intermittent discharge from Glen Canyon Dam.

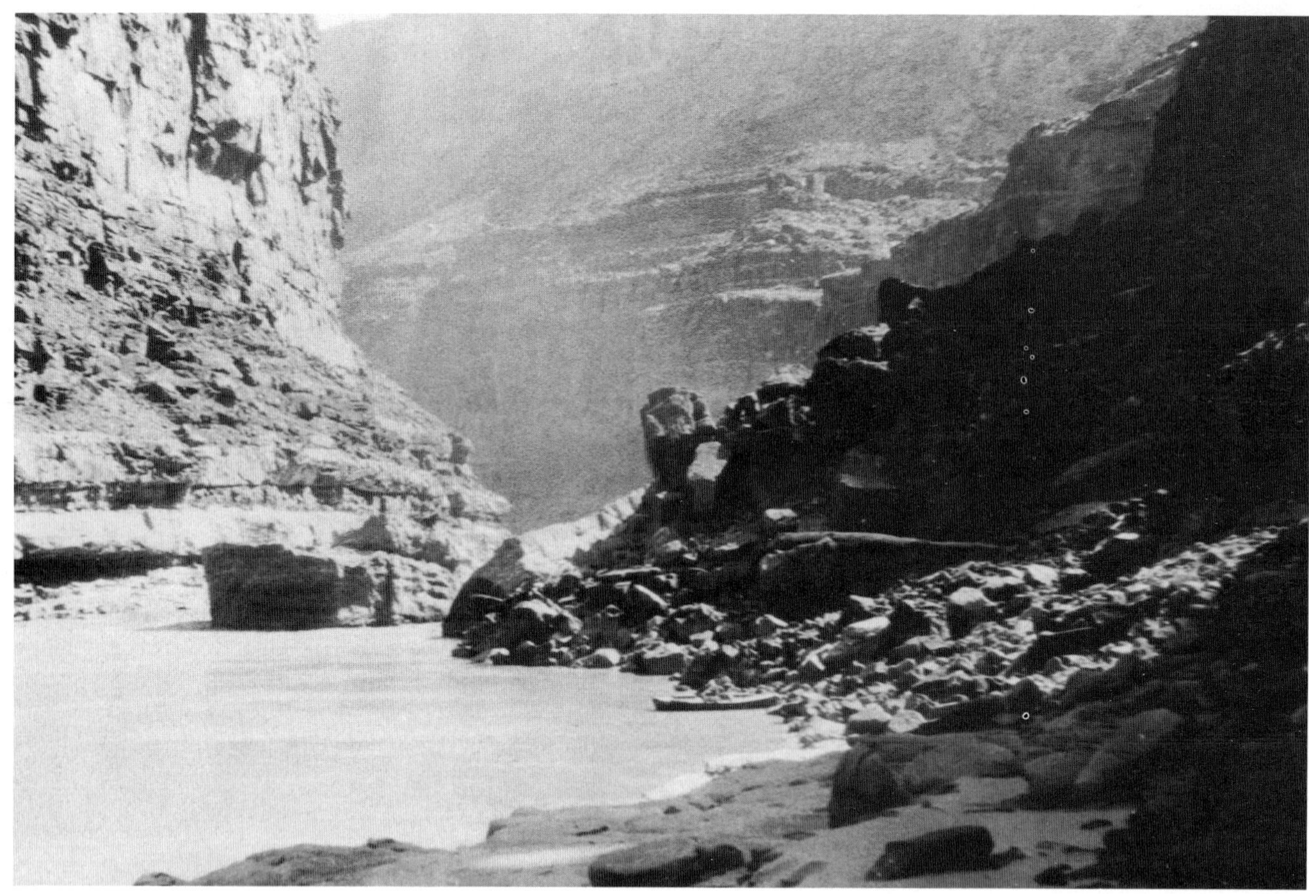

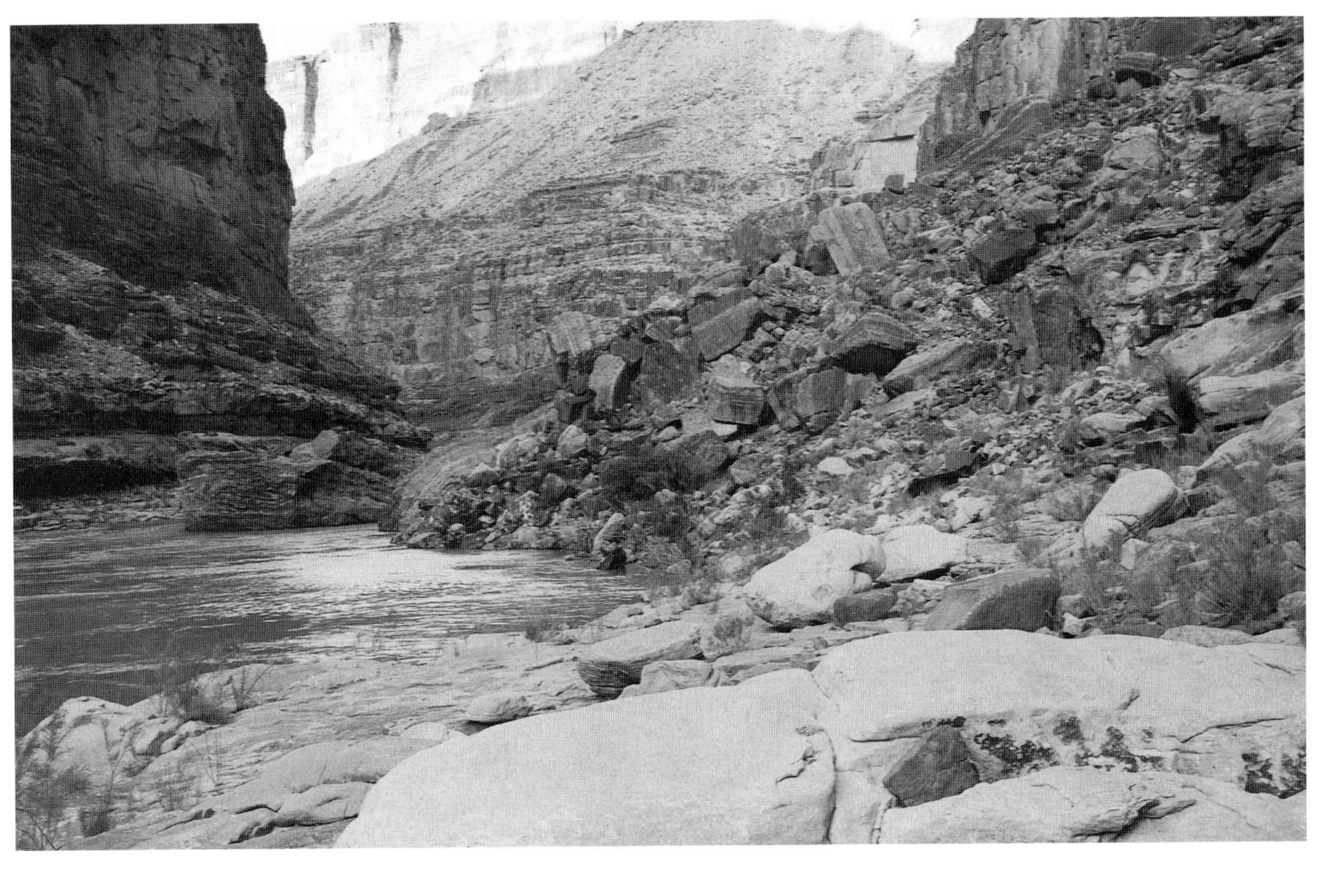

8. House Rock, mile 18.6,
September 9, 1991, 6:00 P.M.

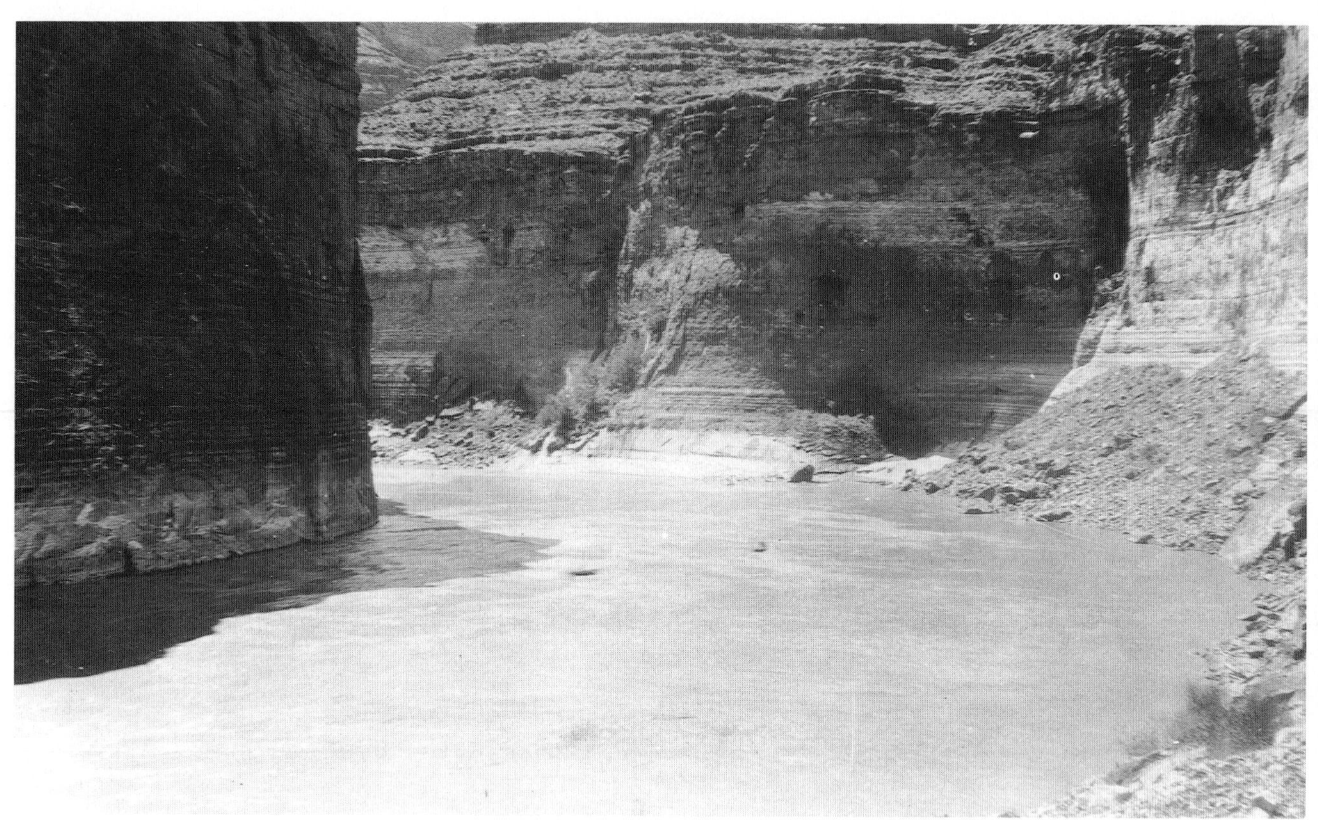

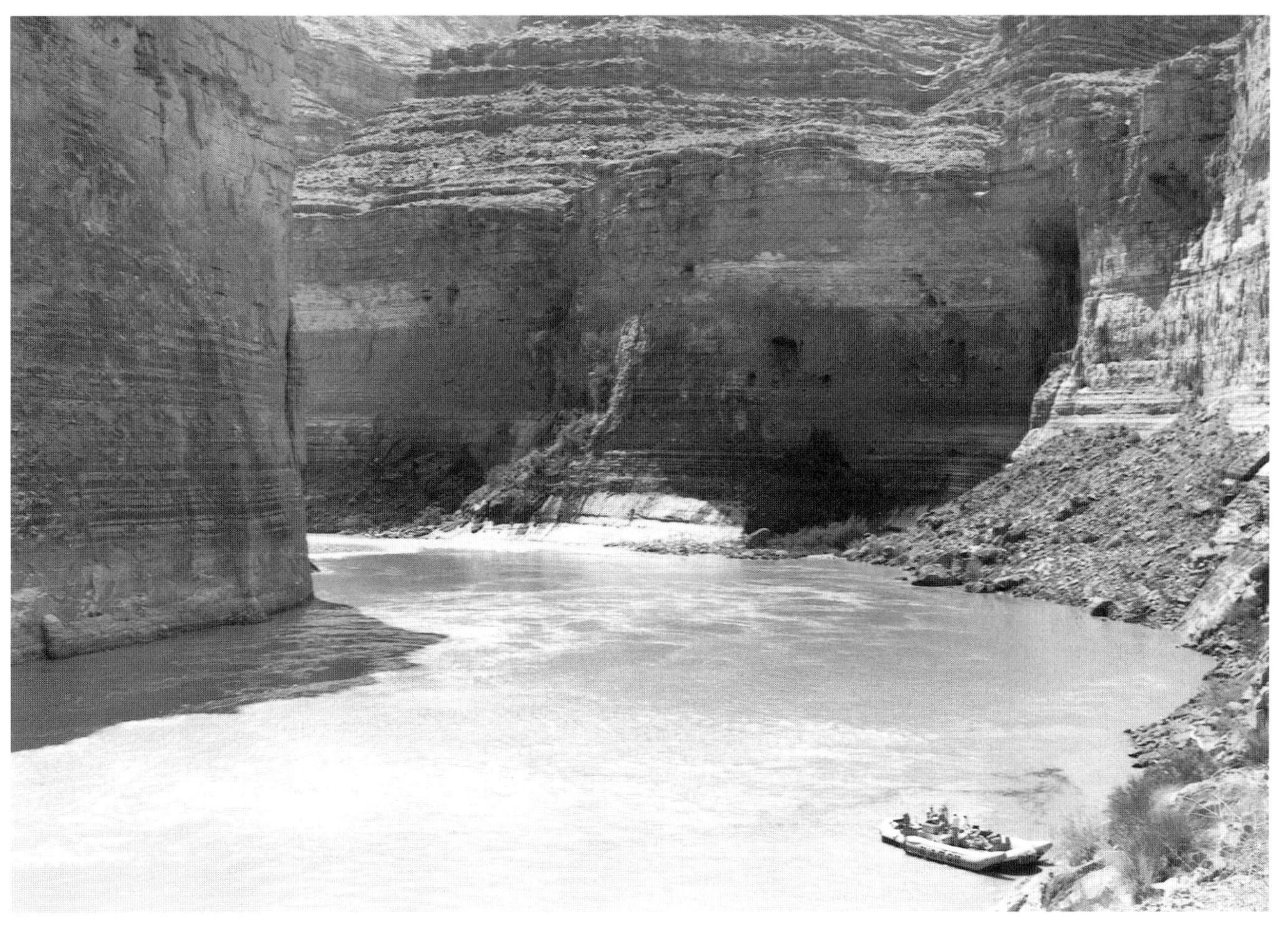

9. Above Vasey's Paradise,
mile 31.6, September 10, 1991,
10:00 A.M.

9. Above Vasey's Paradise, mile 31.6, September 10, 1991, 10:00 A.M.

In a trip down the Colorado, the first view of the spring called Vasey's Paradise, named for G. W. Vasey, a botanist colleague of Powell's, comes as the river curves around a bend to the left near the base of the Redwall Limestone. "On coming nearer," wrote Powell in 1875, "we find fountains bursting from the rock, high overhead, and the spray in the sunshine forms gems which bedeck the wall."

This 1923 photo of an obscured Vasey's Paradise—the spring sits in the shadowed clump of vegetation in the left center of the photos—was taken about mid-morning on August 8 from a ledge of Redwall Limestone about seventy-five feet above the river. The view is from the west side of the canyon, facing almost directly south and downriver.

Several boulders near the river's edge in the center of the 1923 photo are still in place, although the river level is slightly higher in the older photo and the boulders are nearer the water's edge. Tamarisks have taken over the small beach in the low alcove just right of center. Otherwise, there are no discernible changes. Even the three small clumps of vegetation in the lower right corner appear to be the same plants. The two smudges in mid-river in the older photo are probably boats; compare them with the modern boat in the recent photo.

10. Vasey's Paradise from Boulder Island, mile 31.9, September 10, 1991, noon.

Vasey's Paradise is one of the better-known stops in the upper canyon; rafters often halt there to frolic in the warm water from the spring. One of the crew on Powell's first trip, mountaineer John Sumner, called it "the prettiest spring I ever saw" (Cooley 1988:147). Several waterfalls flow from springs emanating from the Redwall Limestone, down over a mat of willows, monkey flowers, and poison oak. This 1923 photograph, taken on August 8 by Freeman, looks back to the west at Vasey's Paradise from a cobble-covered sandbar in the middle of the river, which Birdseye's group called Boulder Island. Lewis Freeman photographed Eugene LaRue taking a photograph with his panoramic camera. Dan Merriam is the picture taker in the recent photo.

As one might expect, the cobbles and boulders on the gravel bar, subjected to scouring when the river reaches higher levels, have been completely rearranged since 1923. Yet the location of the boulder bar remains constant relative to the bend in the river. Like other beach sands along the upper stretches of the canyons, sand that once filled in among the boulders in the foreground has been removed by recent erosion. The river level in the older photo is higher than in the new one.

The flow patterns of the springs appear to be much different in the two photos. A dark stain on the upper left indicates a spring that is not flowing in either photo, although its darker appearance in the 1923 view suggests that it had flowed recently. The second spring from the left is flowing in both photos; its volume appears greater in the recent one but is difficult to judge from the photos alone. The third spring that is highly visible in the 1923 photo seems not to flow at all in the newer shot, although it may be obscured by vegetation. A fourth spring, flowing copiously near the center of the older photo, does not appear to be flowing at all in the recent one, although staining indicates that the spring has produced water in recent times.

This station was the only one in our series that had been previously published in a rephotographic survey: the U.S. Geological Survey repeated it in a 1980 study of vegetation changes along the river (USGS Professional Paper 1132, fig. 41-B, 1980). The 1980 photo shows considerable dead vegetation, mostly redbud trees (Turner and Karpiscak 1980:58–59). Much of the plant life has obviously recovered since that time.

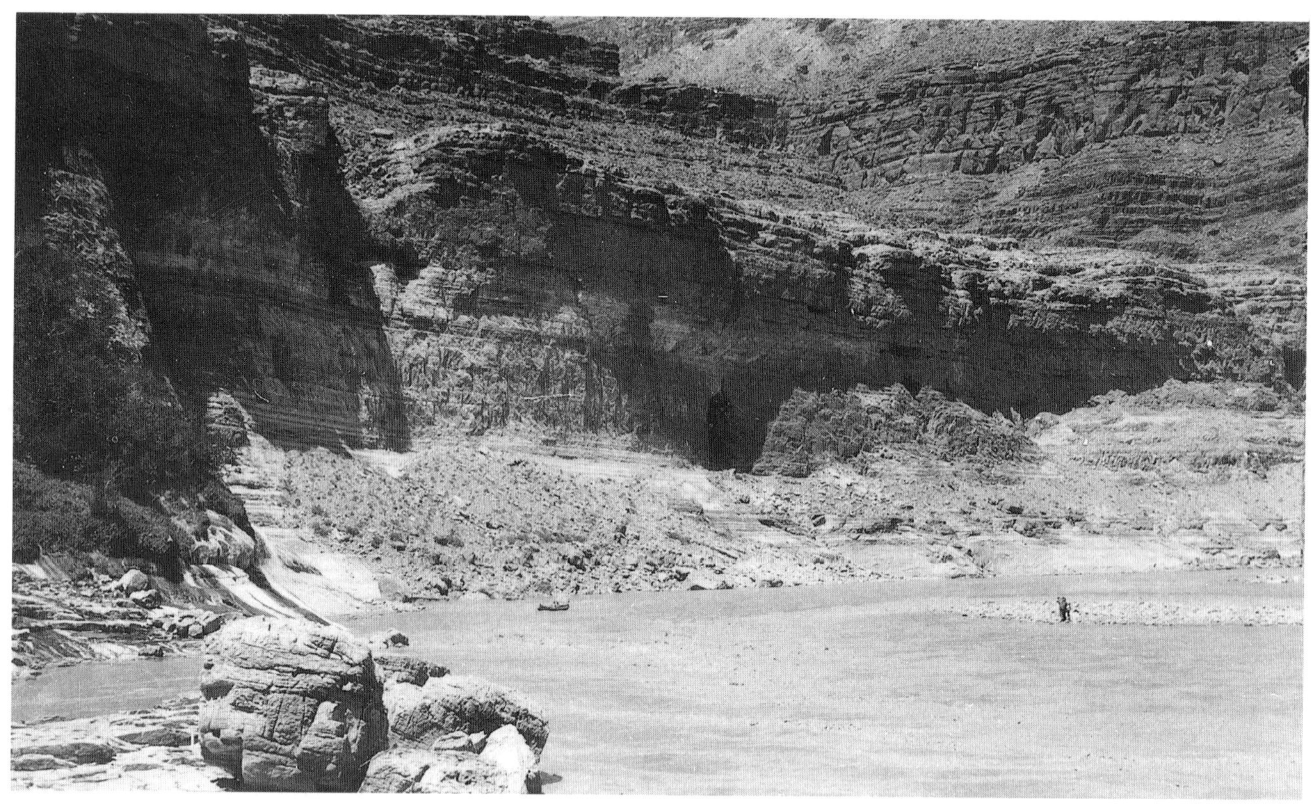

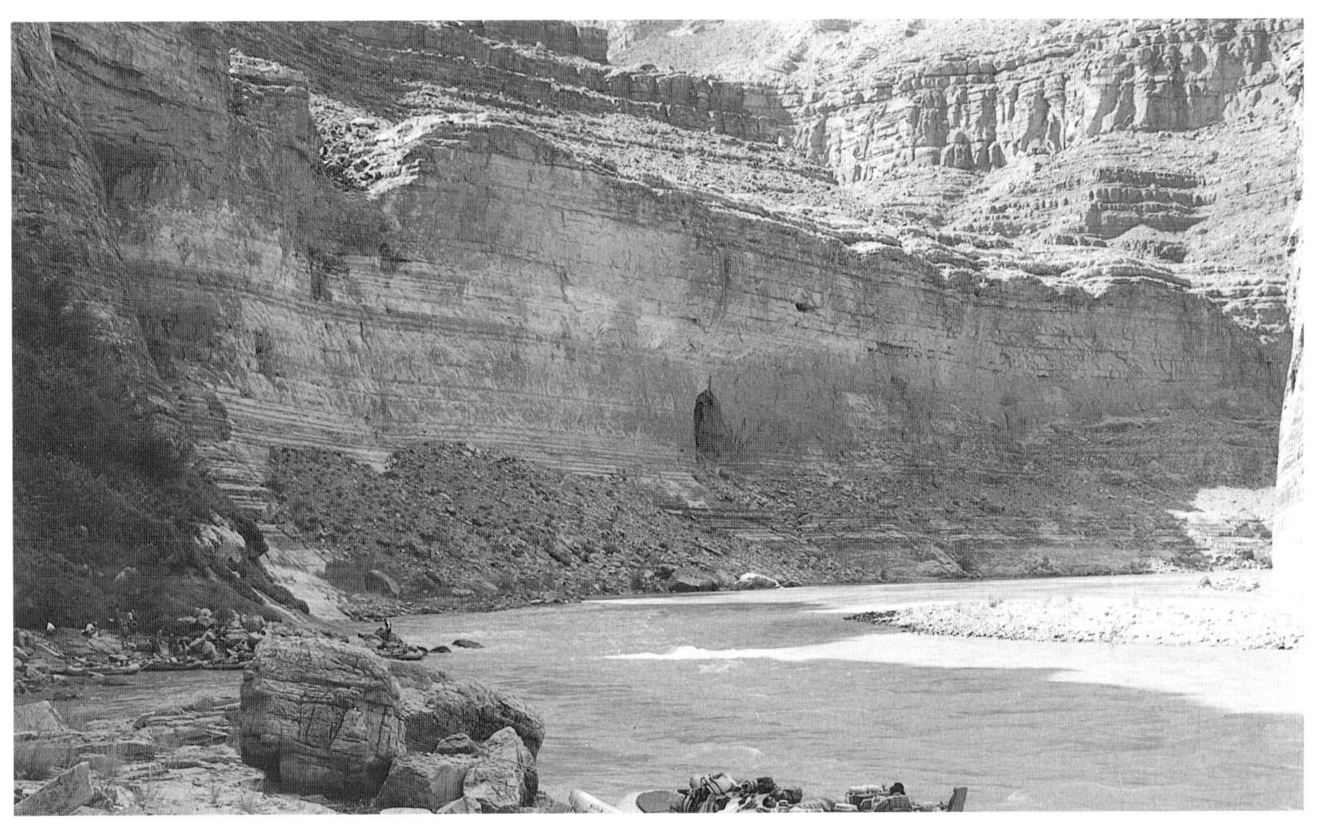

11. Stanton's Cave, mile 32.0, September 10, 1991, 3:00 P.M.

11. Stanton's Cave, mile 32.0, September 10, 1991, 3:00 P.M.

Stanton's Cave is the dark depression in the cliff near the center of these photographs. The terminus of a subterranean drainage channel in the Redwall Limestone, it is named for Robert Brewster Stanton, chief engineer on an 1889–90 expedition to survey a possible railroad route through the canyon from Denver to southern California. Three members of the expedition drowned above this point, and the survey was temporarily abandoned, leaving equipment in the cave for recovery when it continued a few months later. Stanton concluded that construction of a river-level railroad was feasible; obviously, it was never built.

This 1923 photo of Stanton's Cave from Vasey's Paradise was taken at about noon on August 8 by Kolb, looking upstream from the south side of the river. The vantage point is a bench in the lower Redwall Limestone about fifteen feet above river level. In the older photo, Lewis Freeman and E. C. LaRue are barely visible on Boulder Island, taking the previous photograph, along with a boat in the left center of the picture. In the newer photo, tiny figures are savoring spring water below Vasey's Paradise, and the bow of our pontoon raft occupies the extreme lower foreground. Geologically, virtually nothing has changed in this landscape between the dates of the two photographs.

12. Redwall Cavern, mile 33.1, September 10, 1991, 4:00 P.M.

At mile 33 the Colorado River makes a horseshoe bend to the right around a protrusion of Redwall Limestone—a rock layer named by geologist Grove Karl Gilbert, who worked on the Colorado Plateau in the late 1800s. Redwall Cavern, an erosional undercut into the limestone on the left bank, is one of the most recognizable locations on the river. John Wesley Powell estimated, with some exaggeration, that the cavern would seat fifty thousand people.

The 1923 photo in this pair was taken on August 8 from a rock beach at the north edge of the cavern, looking south. In our 1991 photo, Tom McClain stands at the right and Dan Merriam sits on a boulder in the center foreground.

Most of the changes apparent in these photos are on the beach on the cavern floor. The sand in the older photo, taken in early afternoon, is largely untracked, whereas the recent photo shows a beach mottled by the footprints of myriad visitors. Lewis Freeman, in his *National Geographic* account of the 1923 trip, wrote: "The floor was a succession of terraces of hard, smooth sand, rising like the seats of a stadium." Those terraces, formed by water washing up into the cavern, have now been buried by windblown sand.

In the older photo, a line of rocks is clearly discernible on the left side of the cavern floor; most of them are either gone or covered by sand today. On the other hand, several rocks at river's edge, in front of Dan Merriam, are visible only in the new photo. A considerable amount of beach sand beyond the boat in the 1923 picture has since been eroded away.

A rockfall scars the upper cliffs in the far distance of the 1991 shot, beginning in the Kaibab Formation near the rim and strewing rock debris down onto the Esplanade Sandstone. Because it takes several years for newly exposed rocks to oxidize and blend in with surrounding rocks, this spalling must be relatively recent—probably within the past ten or fifteen years. In all the photos we repeated, this is the only one in which a substantial scar from a rockfall appears on a high wall.

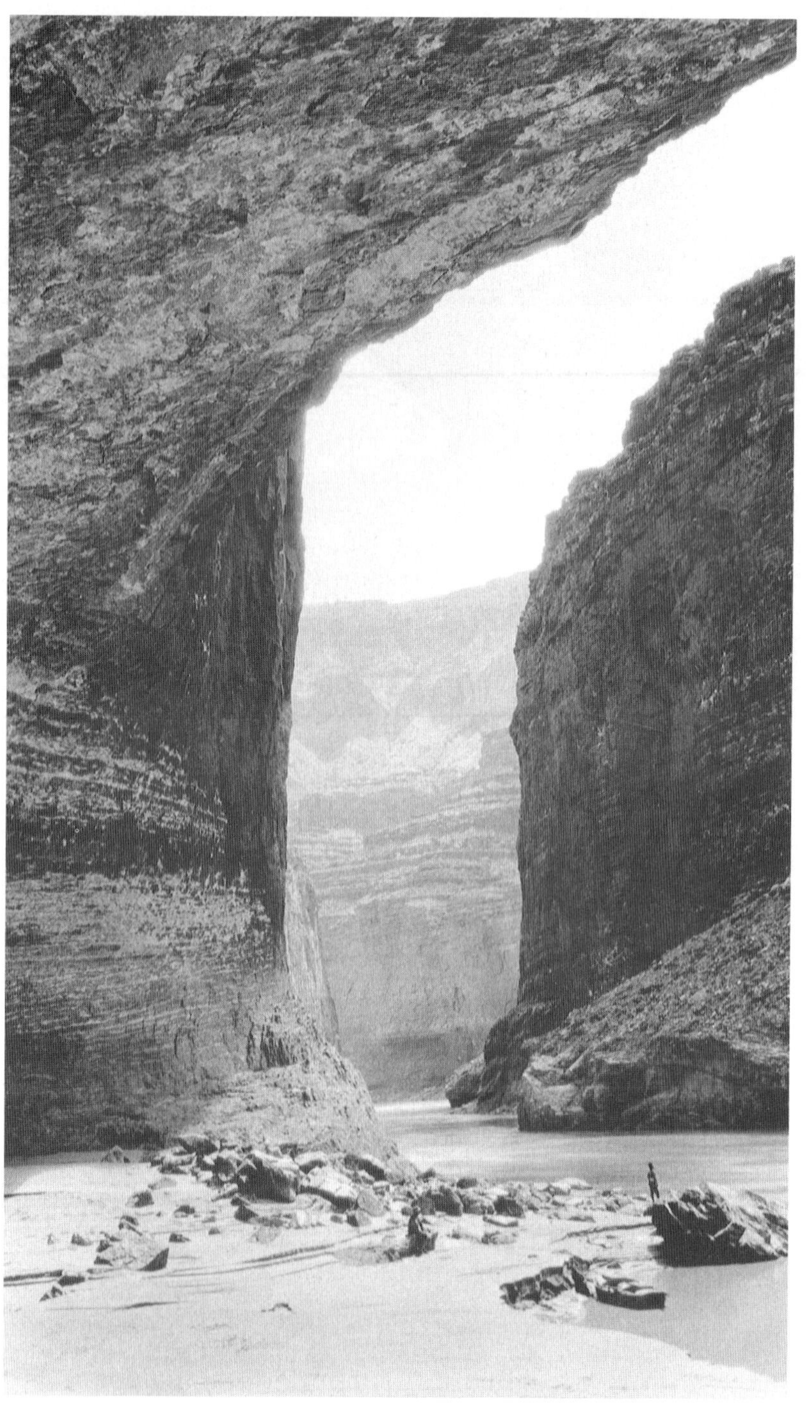

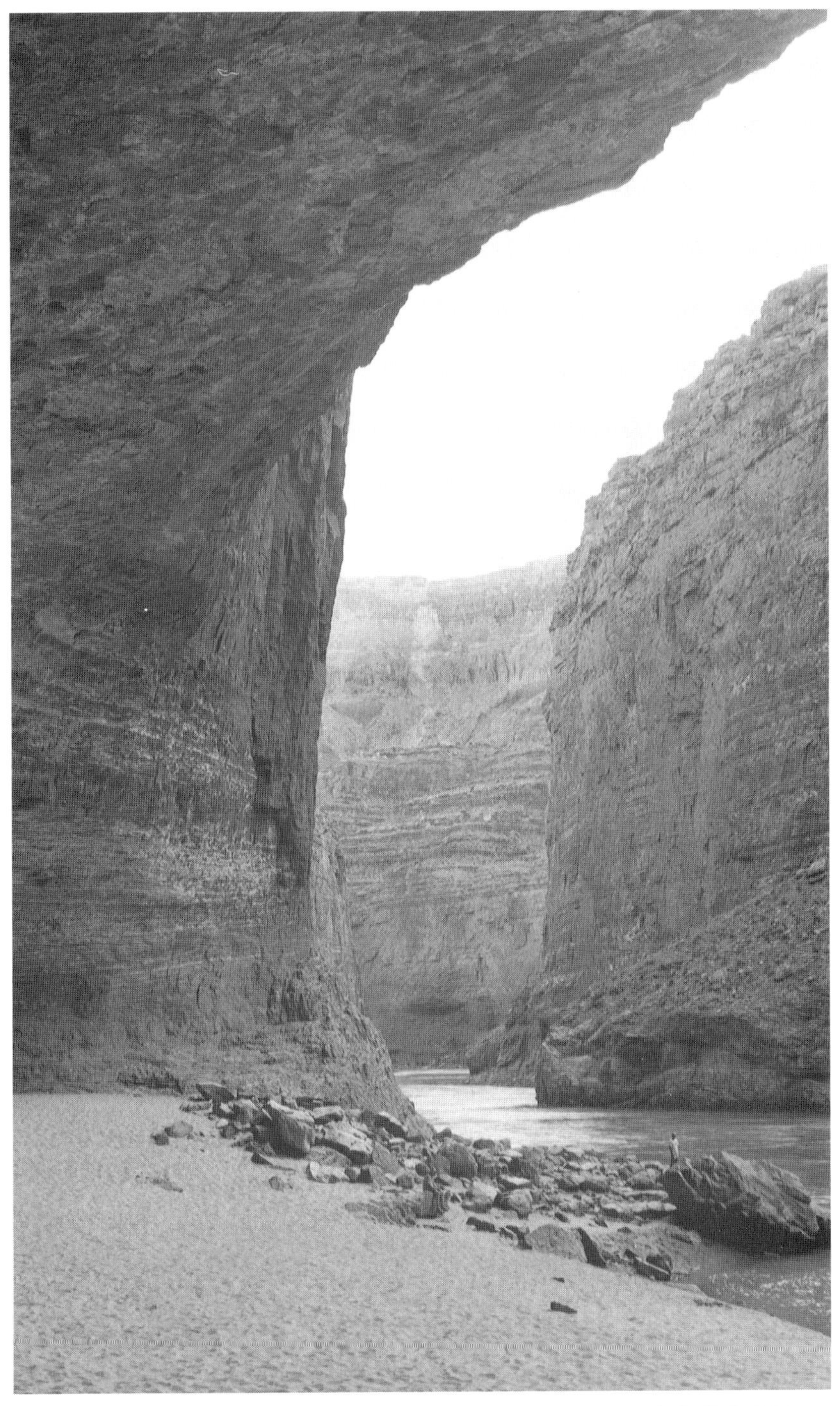

12. Redwall Cavern, mile 33.1, September 10, 1991, 4:00 P.M.

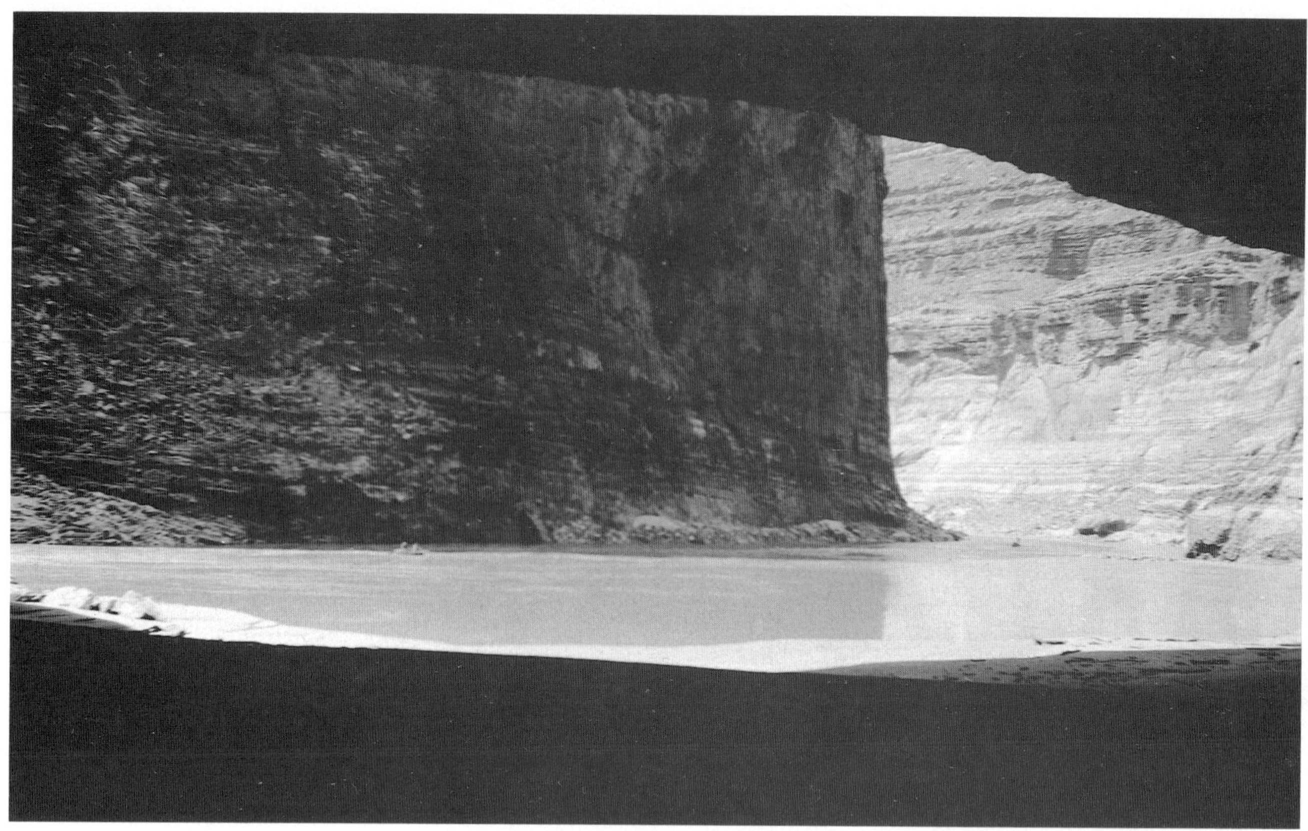

13. Redwall Cavern, view upstream, mile 33.1, September 10, 1991, 5:00 P.M.

This 1923 photo was taken on August 8 by LaRue from the back of Redwall Cavern, looking out of the cave's mouth to the northwest, or upriver. We took the recent shot in late afternoon, apparently only slightly later in the day than the 1923 photo. Shadows make it difficult to compare changes on the near side of the river. The boulder at the middle of the base of the far wall is present in both photos, and there is more vegetation at the base of the far wall in the recent photo. Light-colored features near the shoreline across the river, seemingly boulders, are either absent from the new photo or are obscured by shadows. The boulders and cobbles on the beach at the right are more apparent in the new photo, although that may be a product of lowered river level or the quality of the 1923 photo.

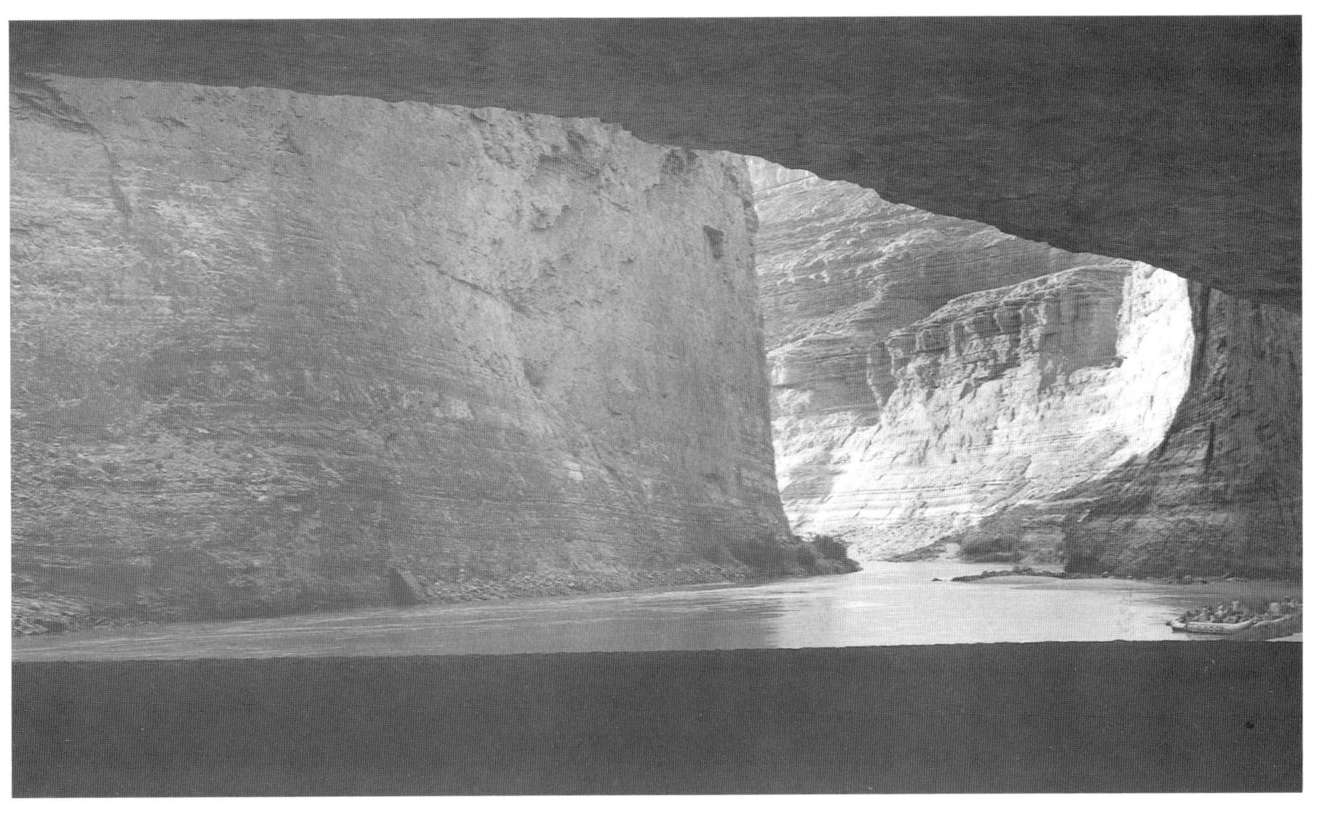

13. Redwall Cavern, view upstream,
mile 33.1, September 10, 1991,
5:00 P.M.

14. Mile 34.4, September 10, 1991, 6:00 P.M.

This photo offers a view 1.4 miles downstream from Redwall Cavern, looking back upstream, or almost directly north. The earlier version was shot on August 8, 1923, by Kolb, from the west bank of the river, about thirty feet above river level on a steep talus slope. We repeated the photo in the dim light of early evening, after which we floated to our camp at President Harding Rapid, nearly ten river miles away, in unnerving darkness.

The talus slope in the foreground is now obscured by a dense stand of tamarisks. The sand spit in the center of the photo, now stripped of most of its sand, has also been colonized by these prolific trees.

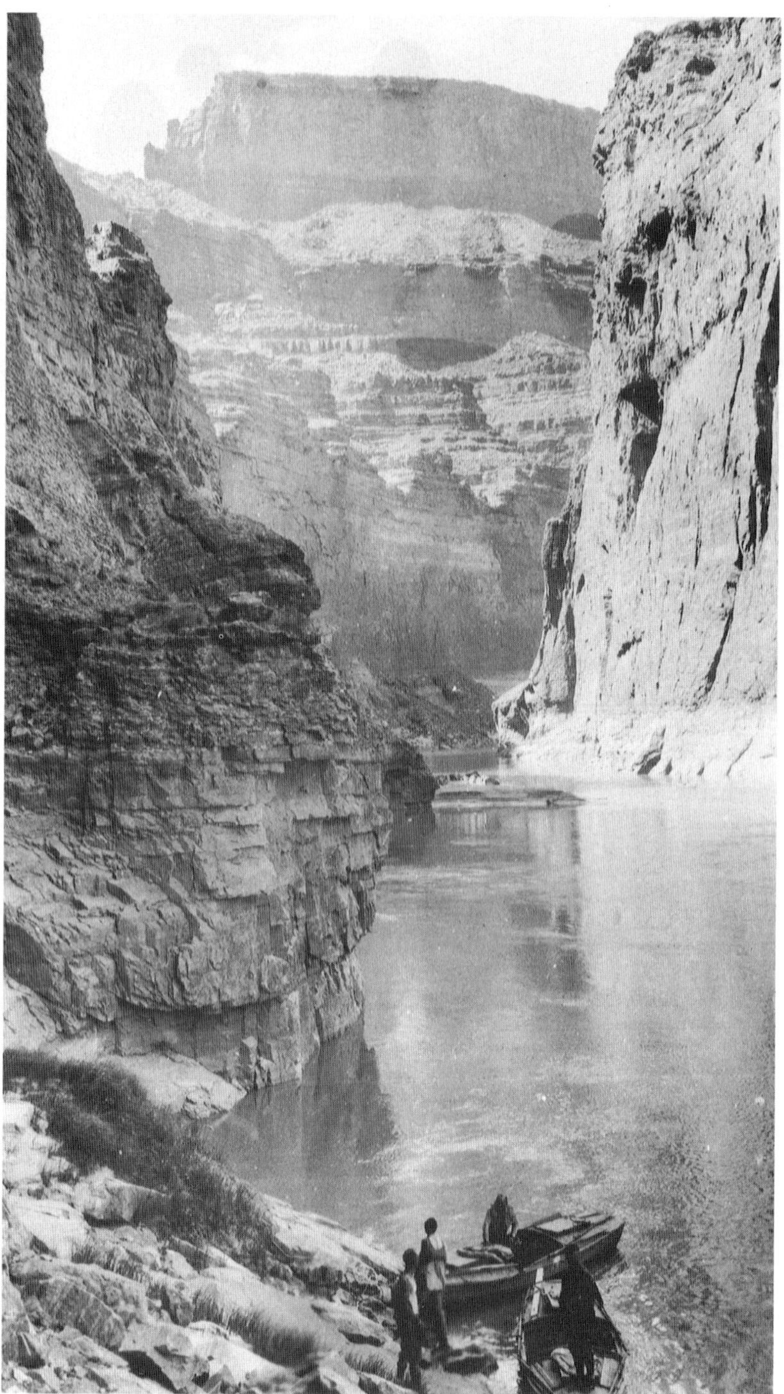

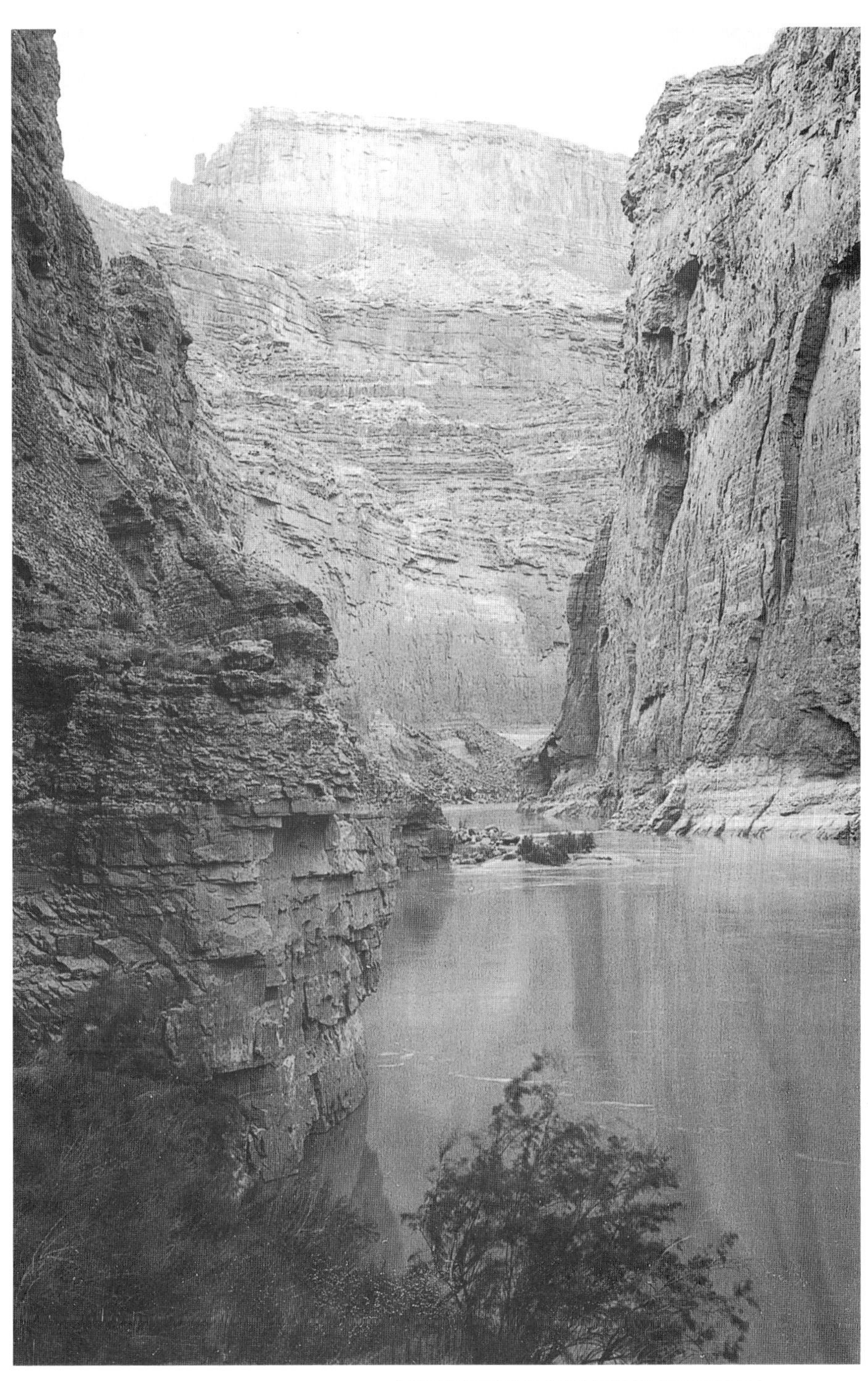

14. Mile 34.4, September 10, 1991, 6:00 P.M.

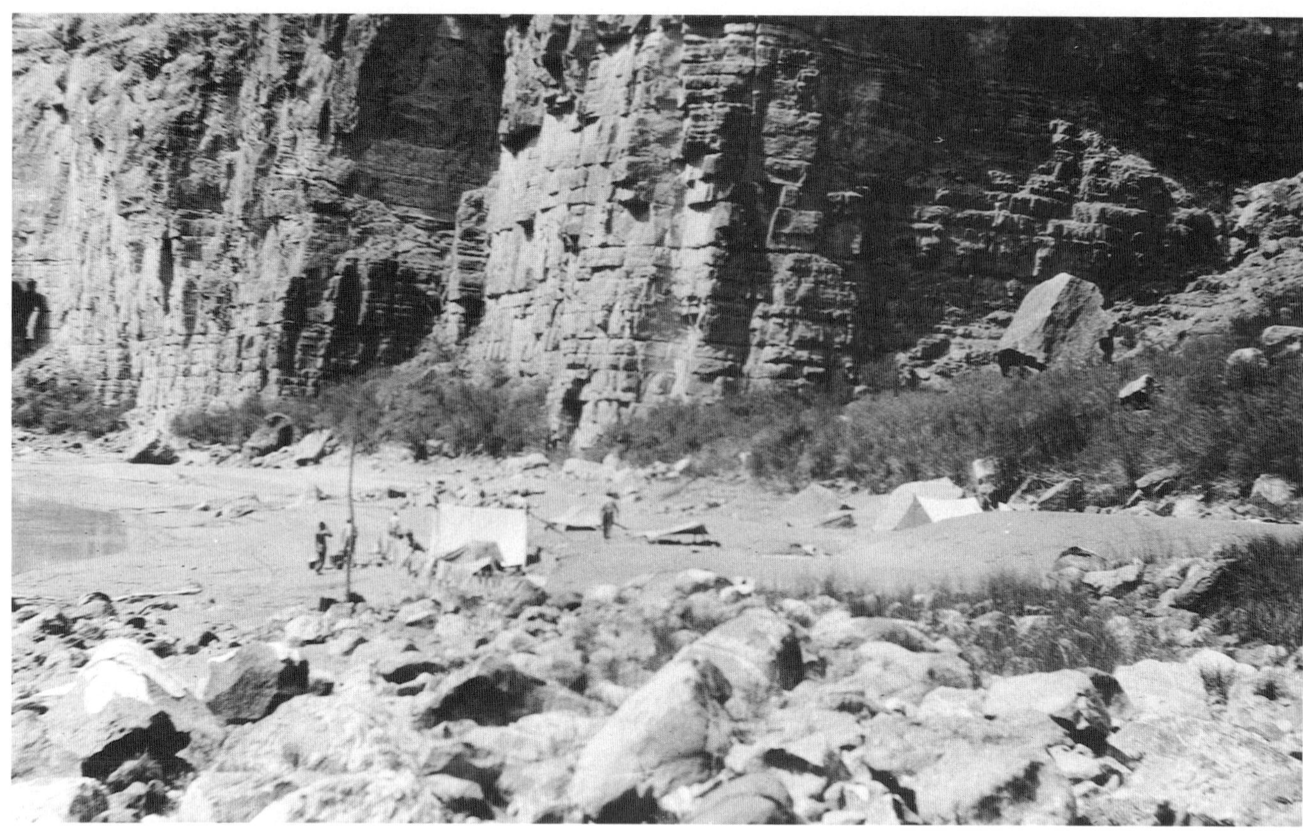

15. Camp at President Harding Rapid, mile 43.6, September 11, 1991, 10:00 A.M.

The 1923 crew laid over here the day of President Harding's funeral and named the rapids after him. They camped on the sandy beach above the rapids, on the east side of the river, spreading their laundry out to dry on some boulders in the left foreground of the older photo.

This photo was probably taken about midday on August 10 by Freeman, south of the camp, looking north-northeast from the edge of a small side canyon that enters from the east. In 1991, we had to choose a camera station a few yards upslope because thick growths of vegetation, mostly tamarisks, have hidden much of the beach. Indeed, vegetation was so dense at this site that we could not repeat a second photo taken nearby. Nevertheless, many of the boulders at the mouth of the side canyon remain visible today—most noticeably the large, square boulder balanced above the beach in the right center of the photograph; a new square boulder is now in place beside it.

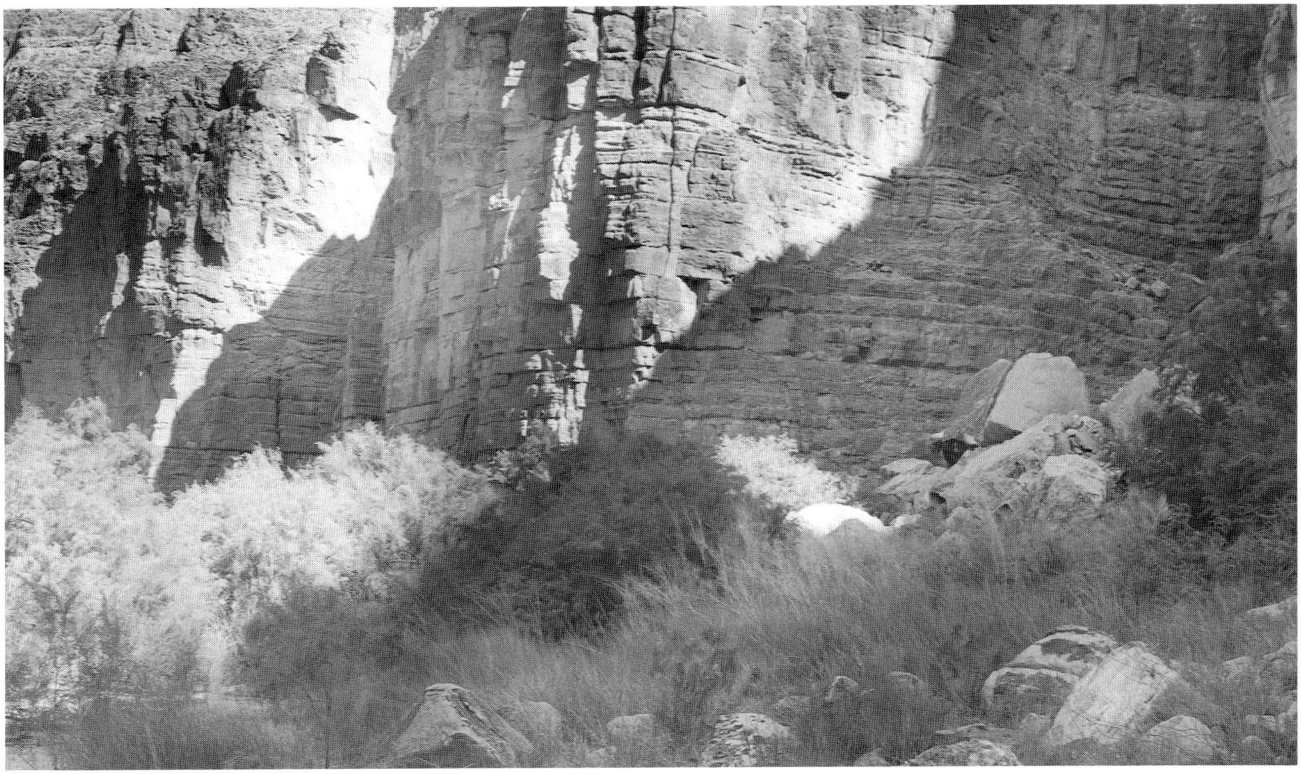

15. Camp at President Harding Rapid, mile 43.6, September 11, 1991, 10:00 A.M.

16. Near Saddle Canyon, mile 46.9, September 11, 1991, 11:30 A.M.

These photos—the earlier version was taken about midday on August 11, 1923—look from the east bank of the river downstream toward Saddle Canyon, which enters from the west. The vantage point is a small sandbar now partially covered by salt cedars. The features at the far edge of the river in the newer photo are two pontoon boats.

Perhaps the most impressive feature of this pair is the remarkable lack of change in the talus slope directly across the river. Both the low angle of the slope and its covering of vegetation, obvious in both photographs, act to retard rock movement down the face of the talus.

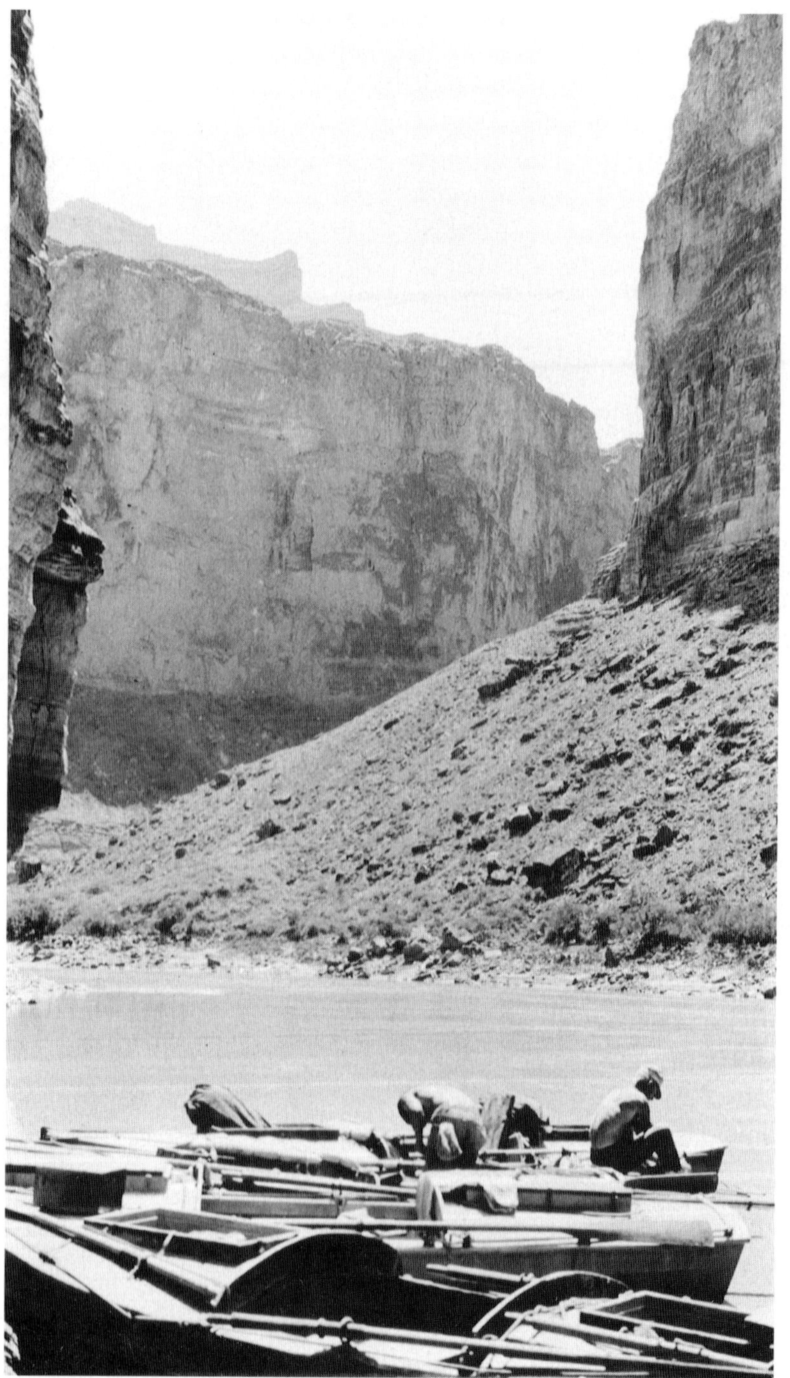

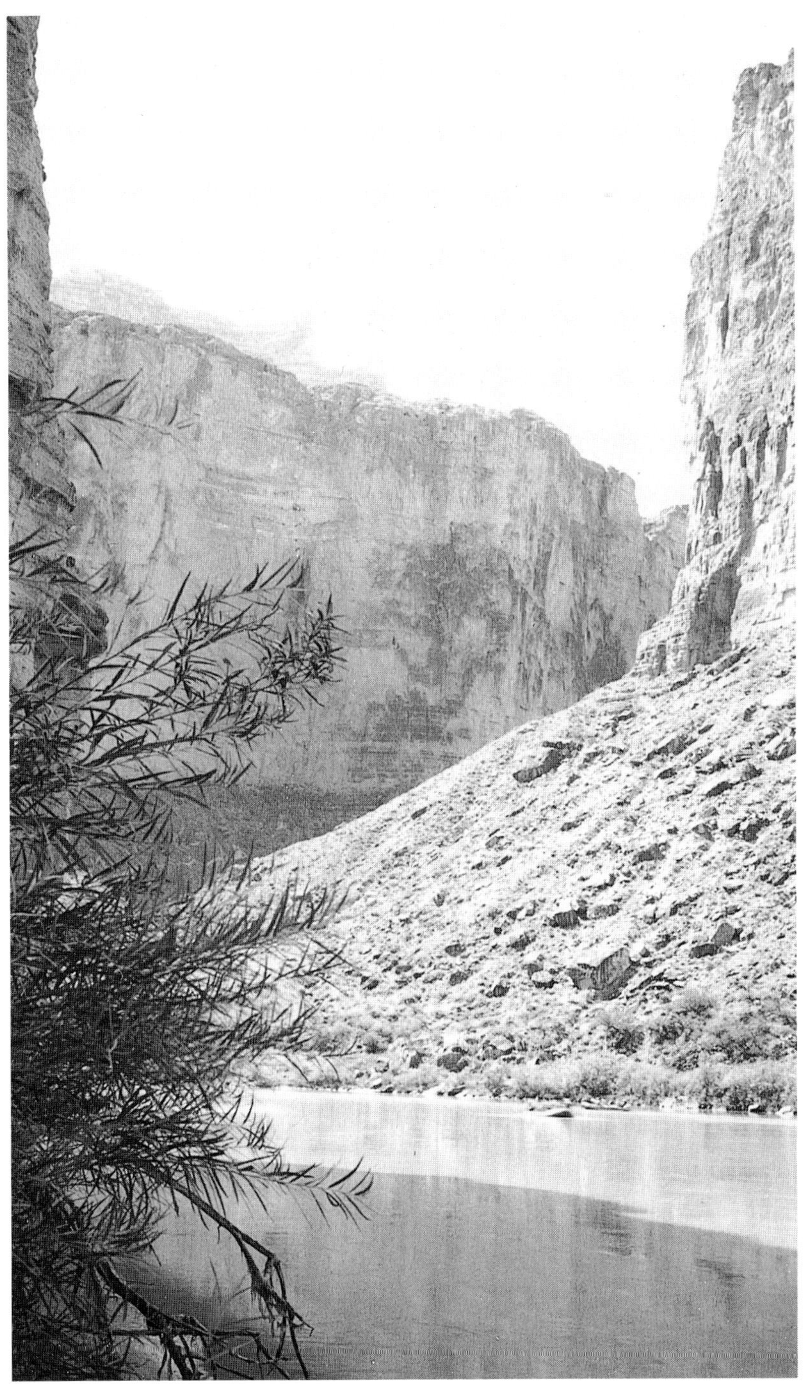

16. Near Saddle Canyon, mile 46.9, September 11, 1991, 11:30 A.M.

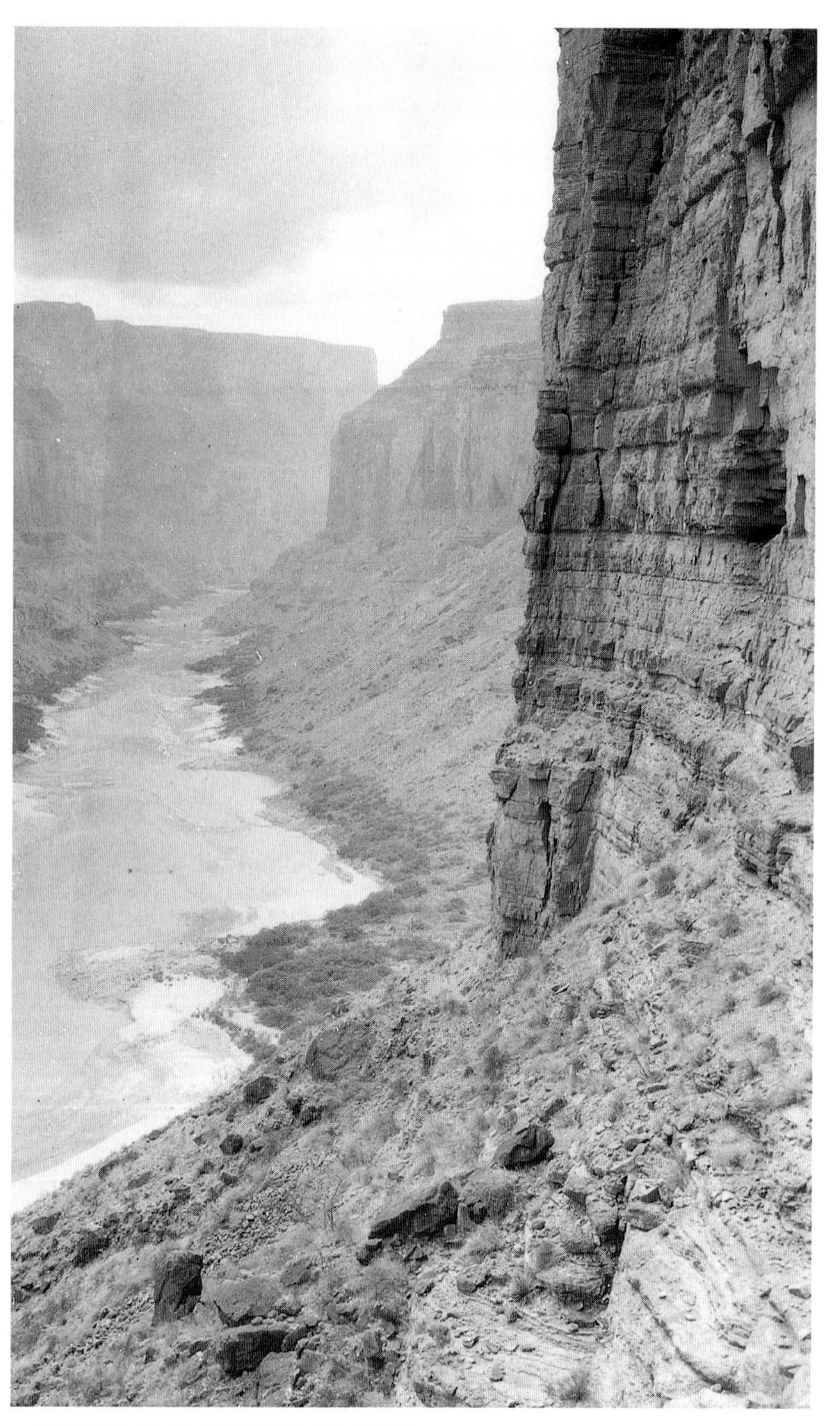

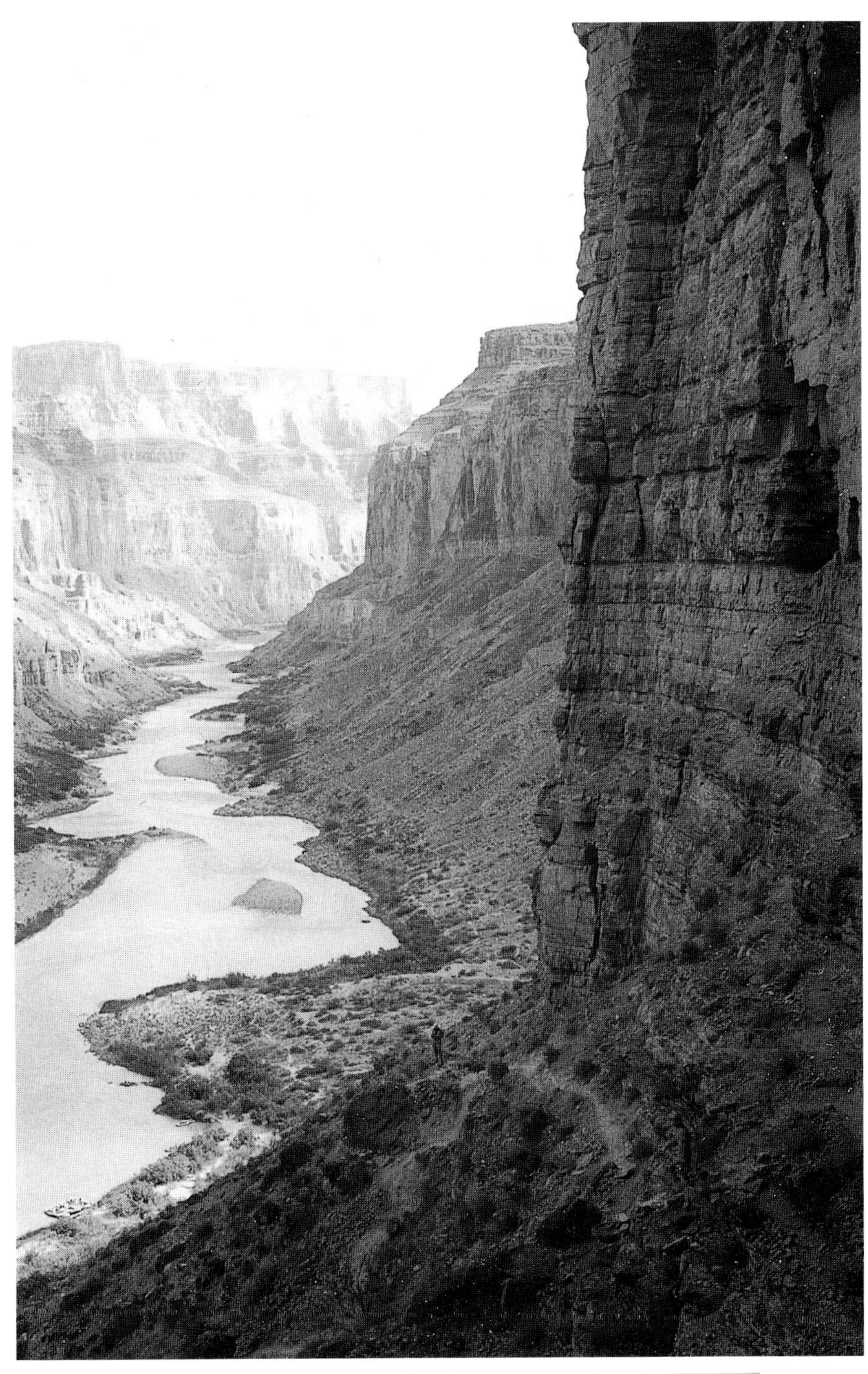

17. Nankoweap Ruins,
mile 53.0,
September 11, 1991,
3:00 P.M.

17. Nankoweap Ruins, mile 53.0, September 11, 1991, 3:00 P.M.

This unique vantage point has been a favorite for viewing lower Marble Canyon for more than a thousand years. About six hundred feet above the river at Nankoweap Canyon are the ruins of several Anasazi granaries. The ancient Puebloans inhabited the Four Corners region and the canyons of the Colorado River, farming the larger tributary deltas, such as the one here at the mouth of Nankoweap Creek, from roughly A.D. 600 to 1200, when they abandoned their canyon dwellings because of drought. This 1923 photo, taken about midday on August 12, shows at the far right one of a series of openings in the cliff face walled in by the Anasazi to provide storage for their seed corn.

River runners regularly visit these ruins, and the 1923 crew made the climb to take several photographs. In his 1986 book, *The River that Flows Uphill,* William Calvin described the view from Nankoweap Ruins: "Looking down the river for four miles, one sees a long, straight stretch of red-walled box canyon, even though the course of the river is serpentine, formed by an alternating series of sandbars and back-eddies with greenery on their edges. A textbook lesson in meanders."

The 1991 photo shows a pronounced increase in vegetation on the nearest beach, along with an increase in the size and extent of sandbars that gives the river even greater sinuousity than it had in 1923. Some of its changing shape may owe itself to today's lower river level, but tamarisks and sandbar willows have also effectively stabilized the shoreline with their root networks, narrowing the channel and fixing its meanders against migration.

A trail marks the base of the cliff in the recent photograph—a result of heavy foot traffic by tourists. A person on the trail and boats (each thirty-three feet long) in the lower left corner of the new photo provide scale.

The vantage point for the 1923 picture had largely eroded away by 1991. E. C. LaRue climbed onto a crumbling talus ledge of shale to take his photo; now, only a steep, loose slope remains. LaRue probably wouldn't use the same site today, but it still looked viable to our boatman, Bill Ellwanger, who climbed up the slope to hold the tripod while this shot was aligned, proofed with Polaroids, and photographed. This may turn out to be the last time this vantage point is ever used. A second 1923 photo of the granary could not be duplicated at all because it was taken from a ledge that no longer exists.

18. Near Kwagunt Rapid, mile 55.8, September 12, 1991, 10:00 A.M.

A few miles below Nankoweap Canyon, the 1923 crew camped on a sandy beach above Kwagunt Rapid on August 12. This photo was taken from the west bank, looking upriver and to the northwest, in the evening as the cook prepared supper and another crew member chopped wood. Elwyn Blake, a boatman on the 1923 trip, wrote in his diary that they "camped at the head of a very rough rapid which, if not run exactly right, will cause serious trouble."

The repeat photograph shows the campsite now overgrown and made unusable by tamarisks—for river travelers, a most frustrating effect of dam construction.

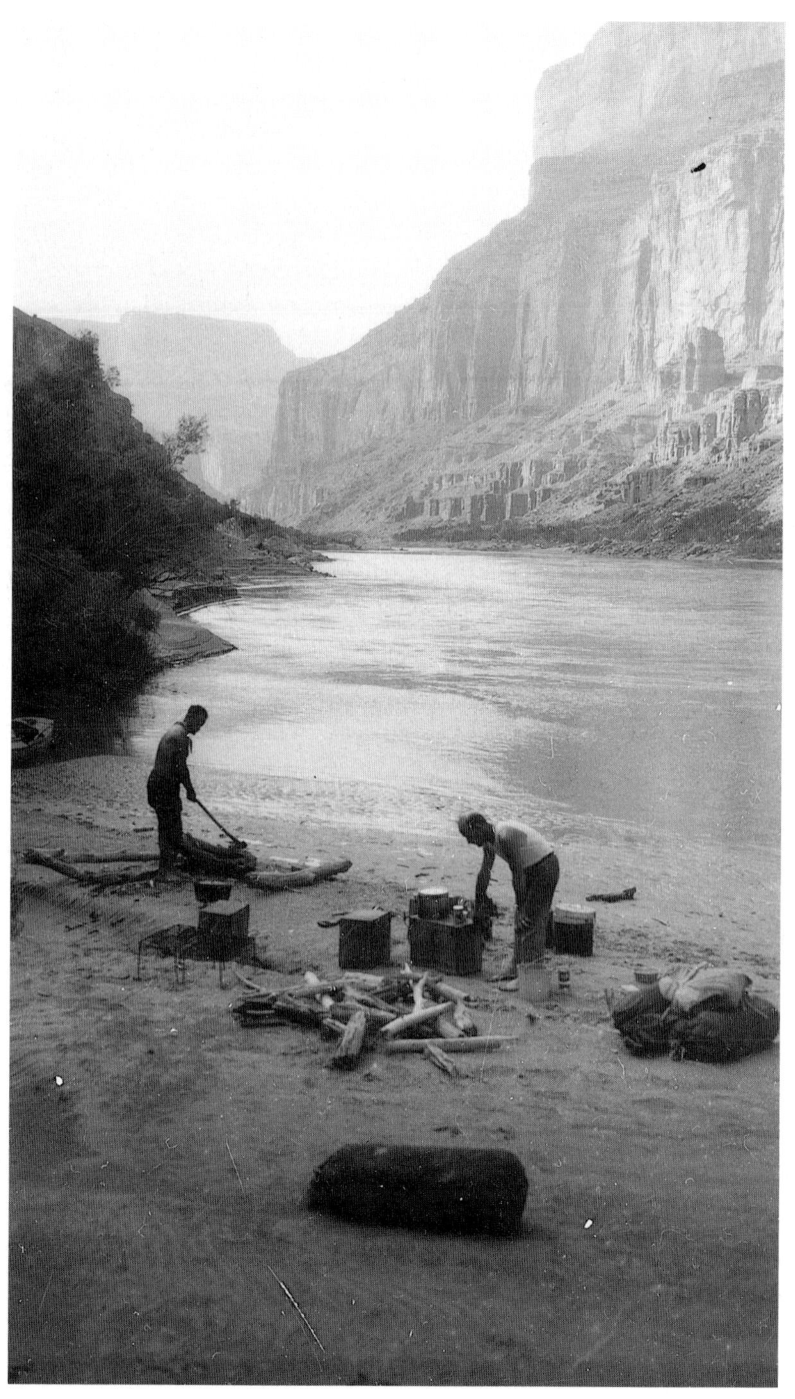

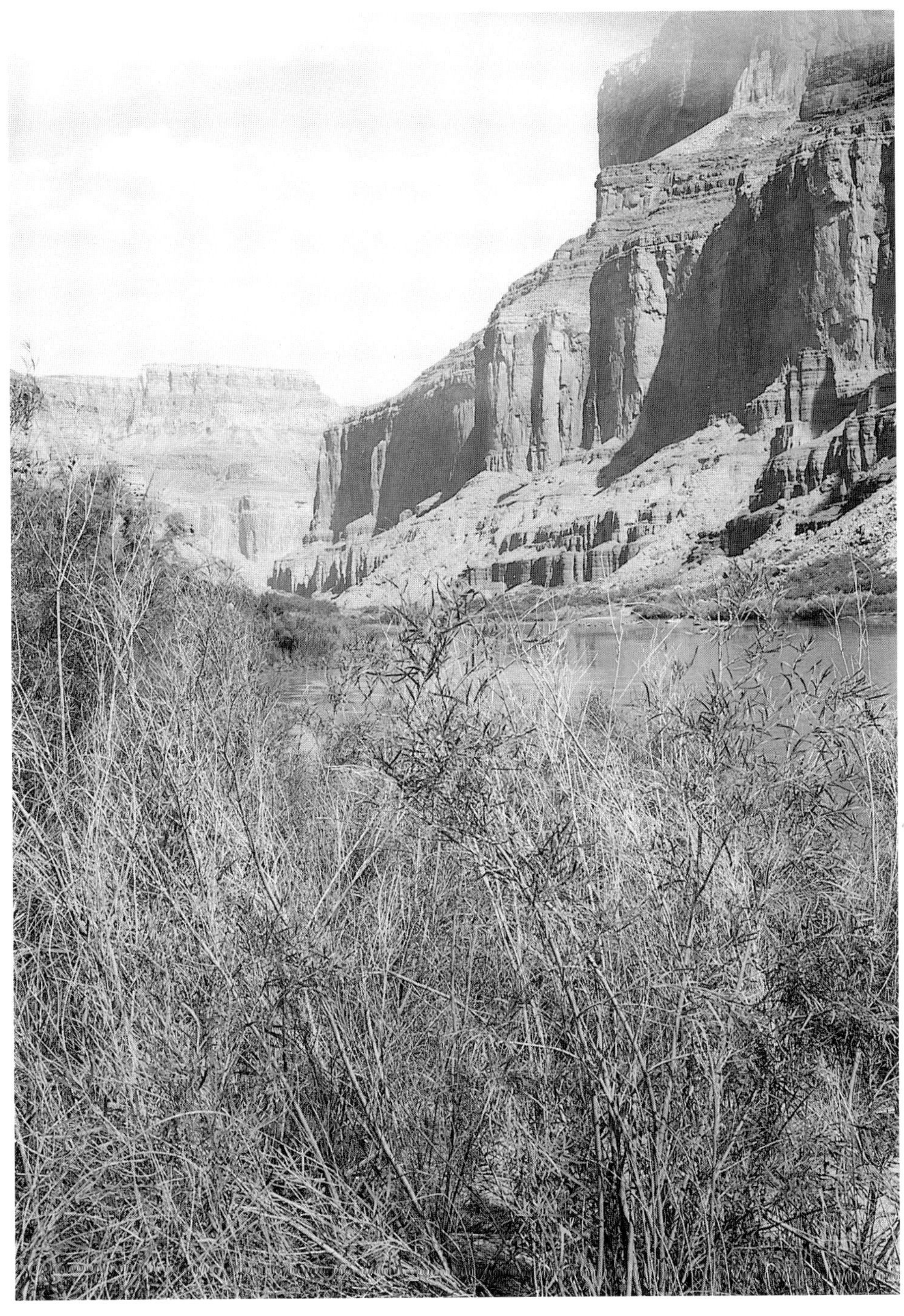

18. Near Kwagunt Rapid, mile 55.8, September 12, 1991, 10:00 A.M.

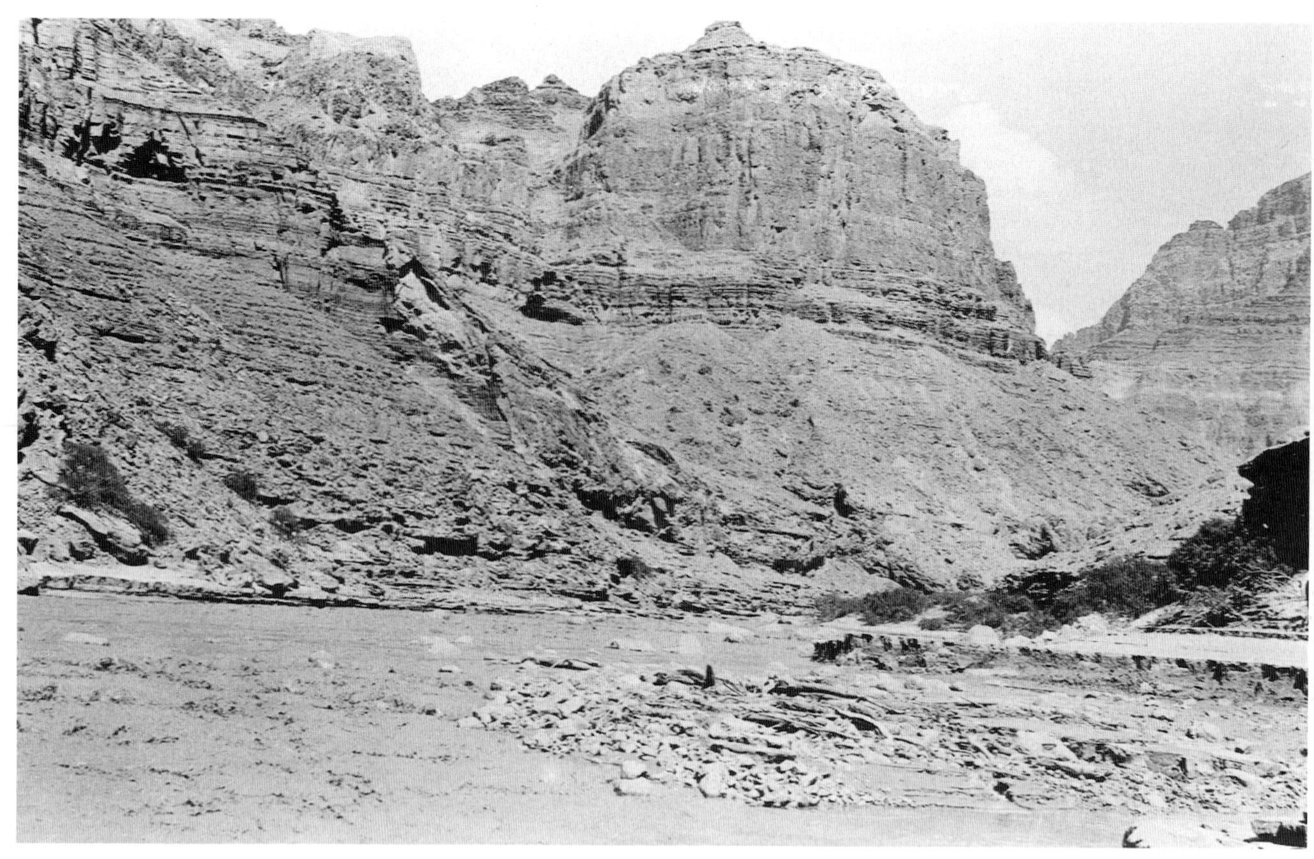

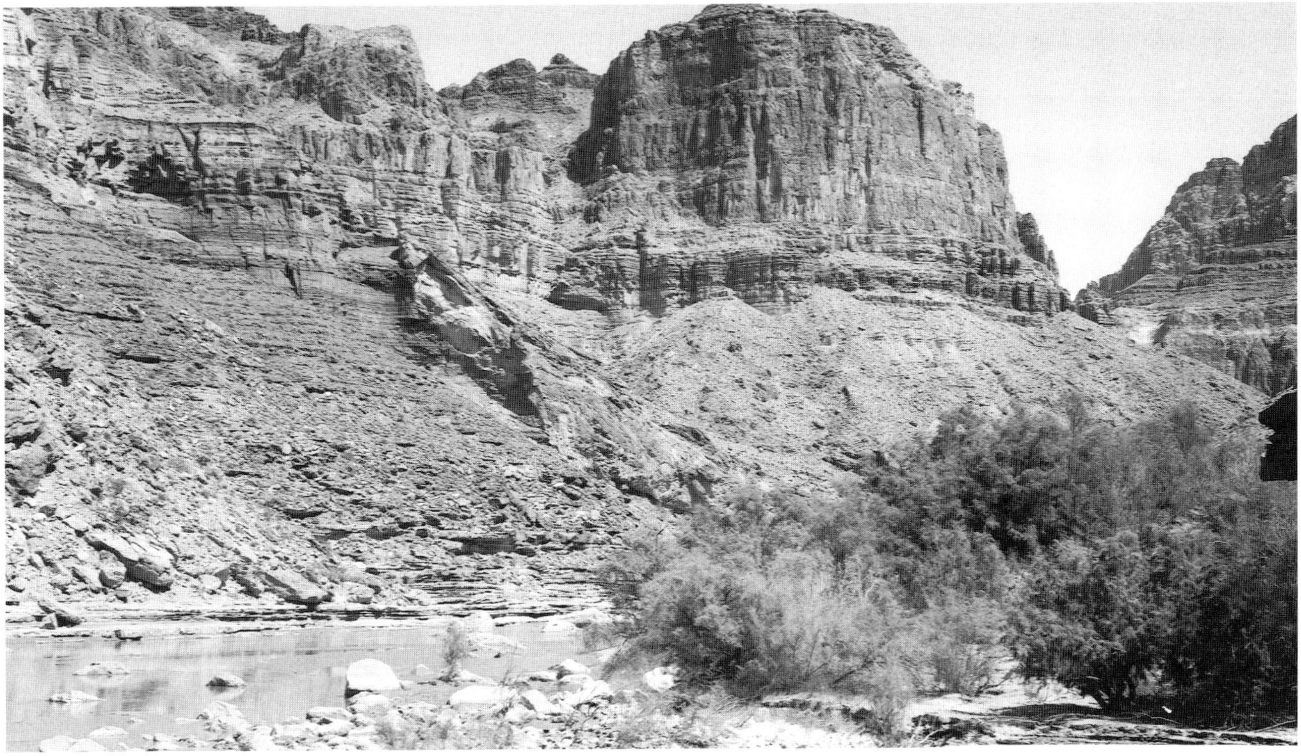

19. The Little Colorado River,
0.5 miles up the Little Colorado
from mile 61.5, September 12, 1991,
noon.

19. The Little Colorado River, 0.5 miles up the Little Colorado from mile 61.5, September 12, 1991, noon.

Below mile 61, the Little Colorado River empties into the Colorado through a narrow defile. There, R. C. Moore drew a detailed cross section of Chuar Butte, across from the confluence of the two rivers. A short hike up the Little Colorado, one of the 1923 photographers took this picture about midday on August 13, looking east-northeast up a tributary canyon from a bank on the south side of the Little Colorado.

Moore's handwritten notation on his scrapbook print of the site, "Looking east near the mouth of Nankoweap," made this vantage point difficult to relocate. The view cannot be seen from the mouth of the Little Colorado, and we recognized it only after passing below the confluence. We landed our boat on the north side of the Colorado and then hiked several hundred yards up the Little Colorado along a bench of Tapeats Sandstone. We took our shot from a ledge of Tapeats; the original photo may have been taken with the camera sitting directly atop a sandstone layer. A ledge of the Tapeats is visible in the lower right corner of both photos.

The photo reveals substantial change along this stretch of the Little Colorado. Besides the usual invasion of tamarisks, many of the rocks forming the riffle in the older photo have been rearranged. The driftwood, unsurprisingly, is now gone, and a rock perched at river's edge on the extreme left of the older photo has been washed away. The talus slopes are noticeably unchanged. The amorphous mass of rock inclined down the talus slope in the center of the view was at some time in the past cemented together by travertine but is now being destroyed by erosion.

20. The Little Colorado River, about 0.5 miles up the Little Colorado from mile 61.5, September 12, 1991, 1:30 P.M.

This second photograph of the canyon of the Little Colorado was originally taken several hundred feet east, or upstream, from the first photo at midday on August 13, 1923. The view is almost directly east. As with the previous shot, Moore was hopelessly lost when he captioned these photos for his scrapbook and labeled this scene "Lower part of Marble Canyon"—another obstacle to site recognition.

Here the bank, rocks, and even the vegetation along the Little Colorado are encrusted with white, crystalline calcium carbonate from springwater that feeds the stream. The largest spring, known as the *sipapu* by Hopi Indians, is believed to be the opening in Mother Earth from which the Holy People entered the present world. It issues mineral-laden waters that form a bright blue stream year-round.

The Little Colorado, depending on weather conditions in its drainage basin, can be very silty, as it is in the 1923 photo. Stopping here with the Powell expedition of 1869, John Sumner called the Little Colorado "as disgusting a stream as there is on the continent. . . . Half of its volume and 2/3 of its weight is mud and silt" (Cooley 1988:148). At other times the water is a murky sky blue because of the scattering of light by its mineral content.

Despite the river's being higher in the older photo, comparison of these two reveals that even some of the larger boulders in the riffle in the original view have since been moved. Vegetation in the 1923 photo appears to be largely mesquite scattered over the bank on the right side of the photo. This bank is now solidly overgrown with tamarisks, and a tamarisk limb frames the upper right of the newer composition. The dark object in the river on the left side of the 1923 image appears to be the head of a swimmer.

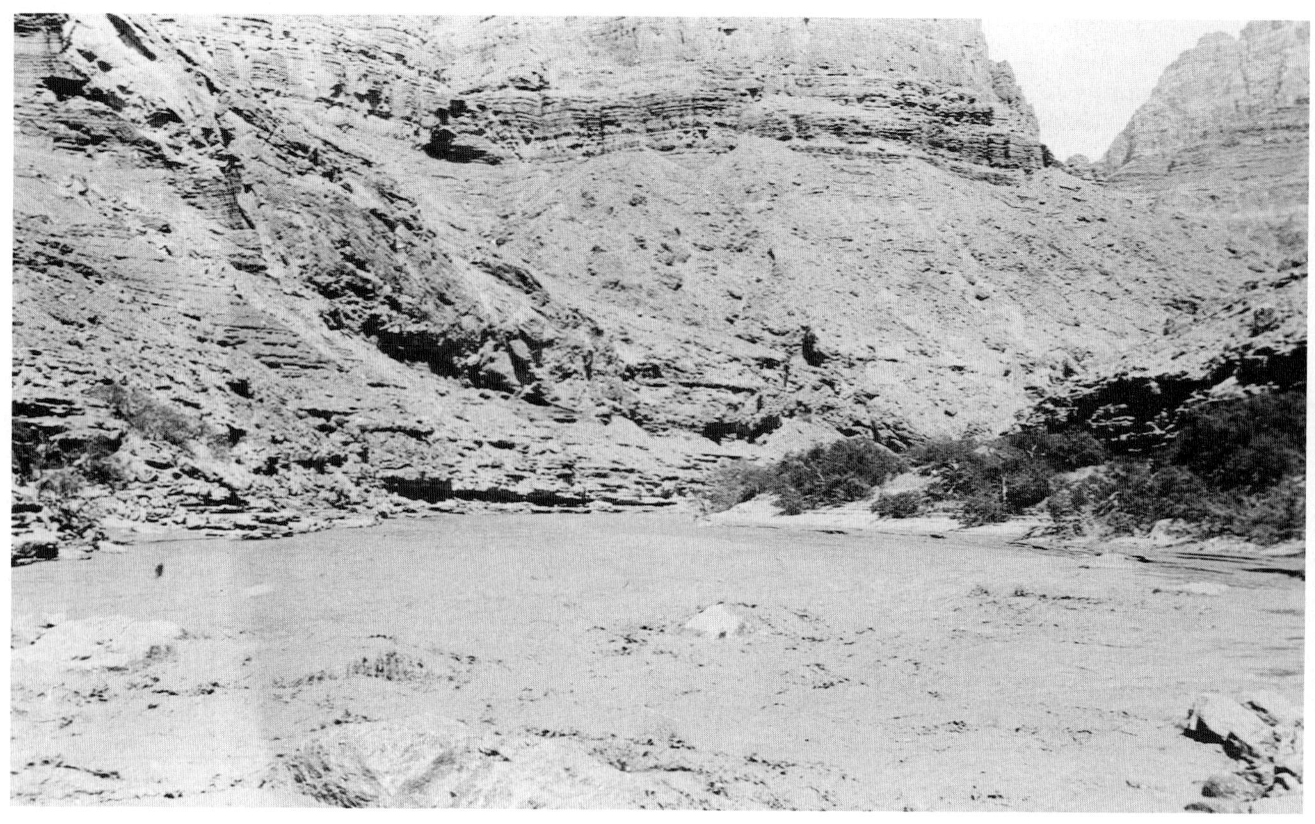

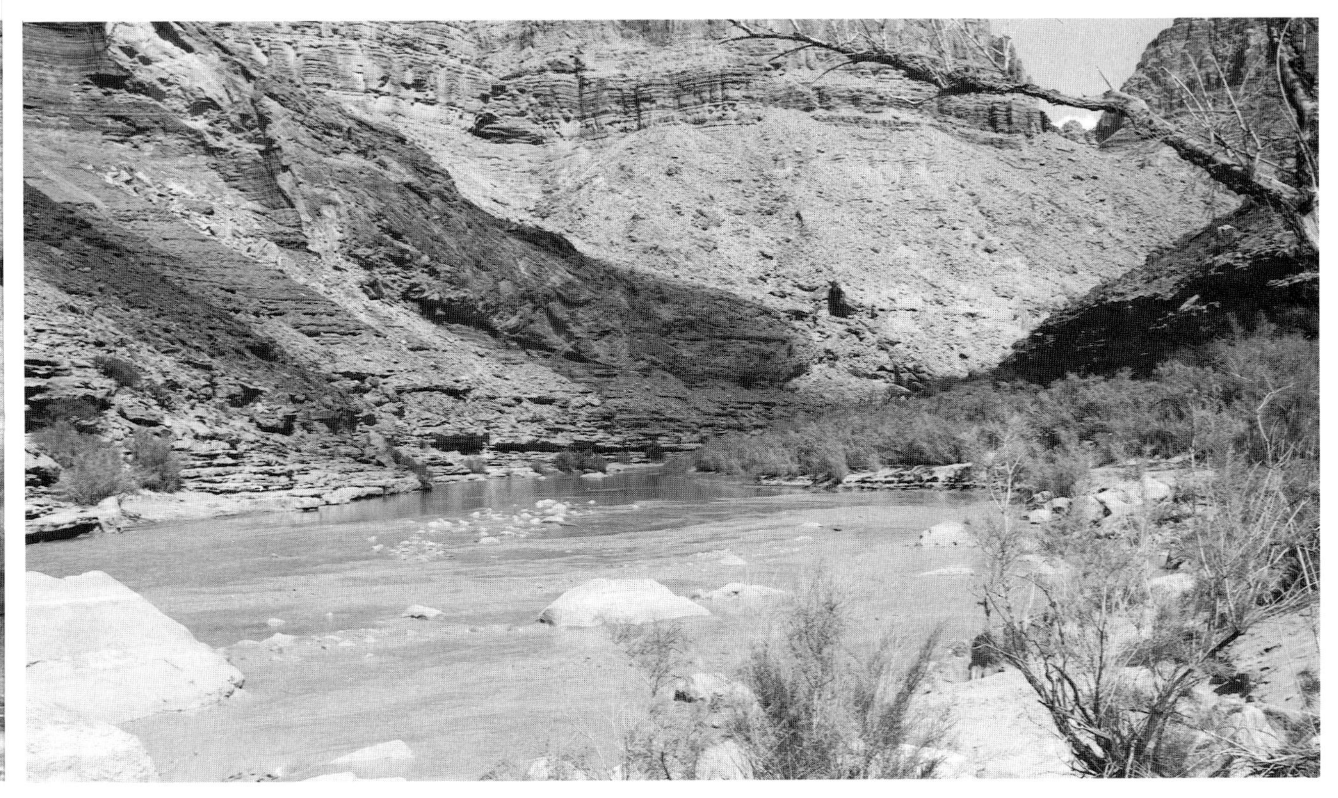

20. The Little Colorado River, about 0.5 miles up the Little Colorado from mile 61.5, September 12, 1991, 1:30 P.M.

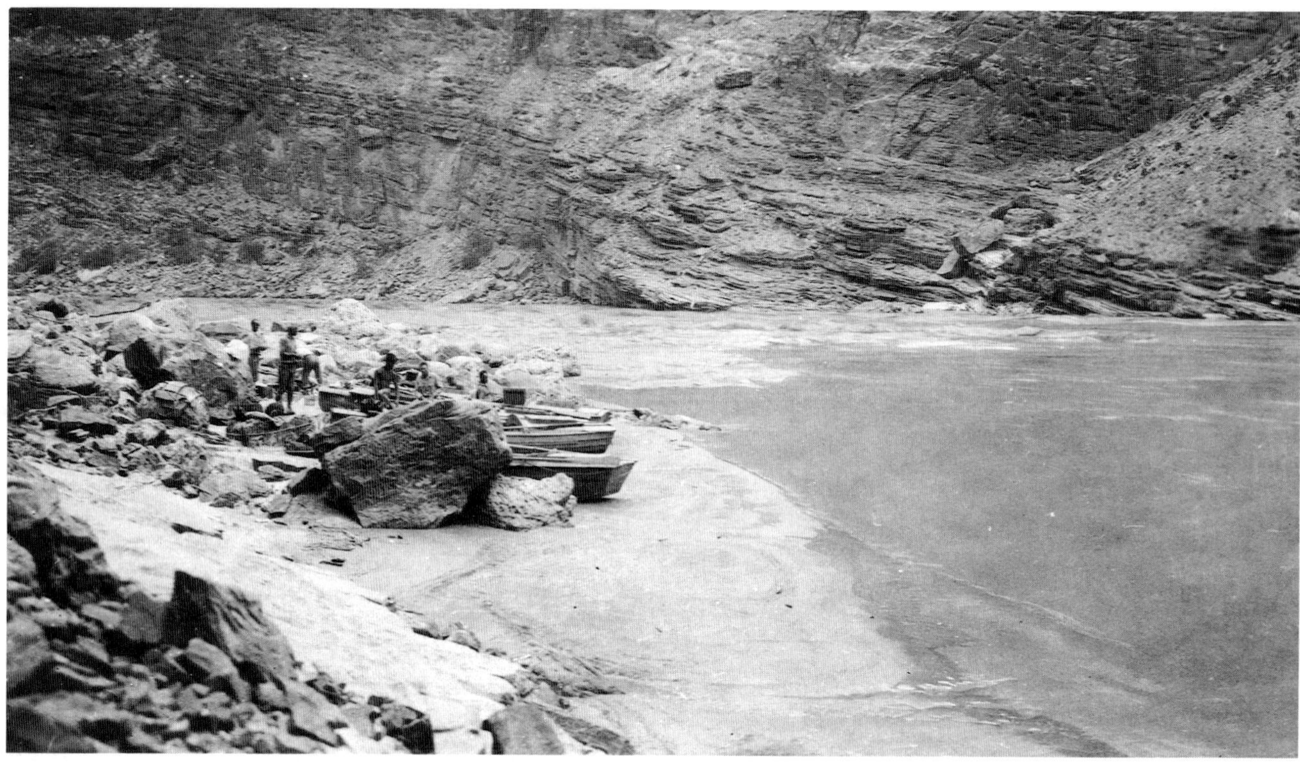

22. Nevills Rapid, mile 75.3, September 13, 1991, 1:00 P.M.

About fourteen miles below the confluence of the Colorado and the Little Colorado is Nevills Rapid, also known as 75-Mile Rapid. It was named for pioneer commercial river runner Norman D. Nevills of Mexican Hat, Utah, whose first paying venture brought botanists Elzada Clover and Lois Jotter down the river in 1938—the first women to complete a river trip through Grand Canyon.

Birdseye's group camped here on the evening of August 15, 1923. Elwyn Blake described the following morning in his diary: "The boats were high and dry this morning as the river dropped and the water where the boats were moored was shallow. Some of the largest waves yet seen were encountered in the rapids below camp." The older photo, by LaRue, apparently taken early on the morning of August 16 as the crew surveyed their situation, looks from the east bank almost due west across the river. Another 1923 photograph shows LaRue in the process of shooting this photo from atop a large boulder.

We took our repeat photograph from the same angle as the original photo, but LaRue's old location is now overgrown with vegetation. Our repositioned site lay on a direct line from his station toward rocks at the head of the rapids. As a result, the foreground in the two photographs differs dramatically. Both photos show the dark, layered Shinumo Quartzite that makes its first appearance a short distance upstream from here.

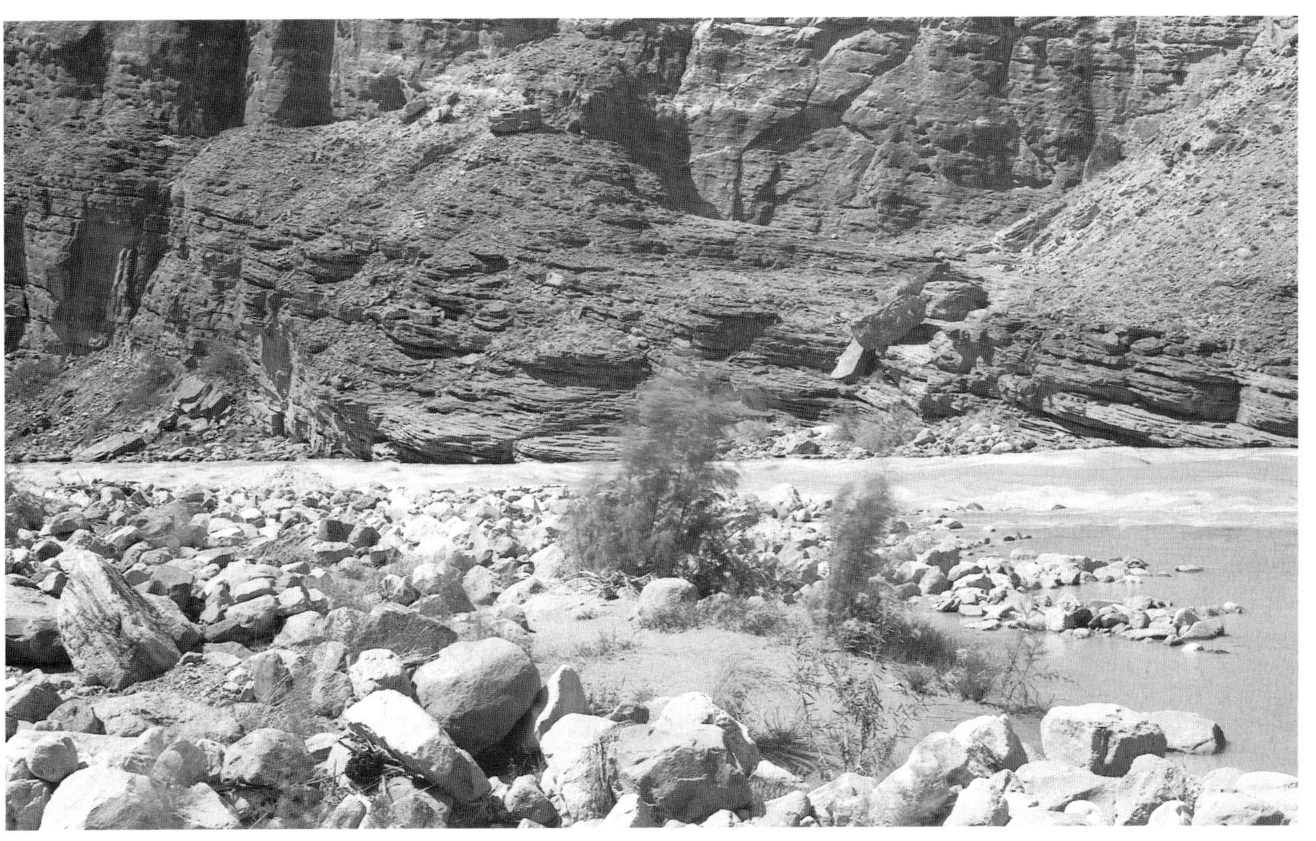

22. Nevills Rapid, mile 75.3,
September 13, 1991, 1:00 P.M.

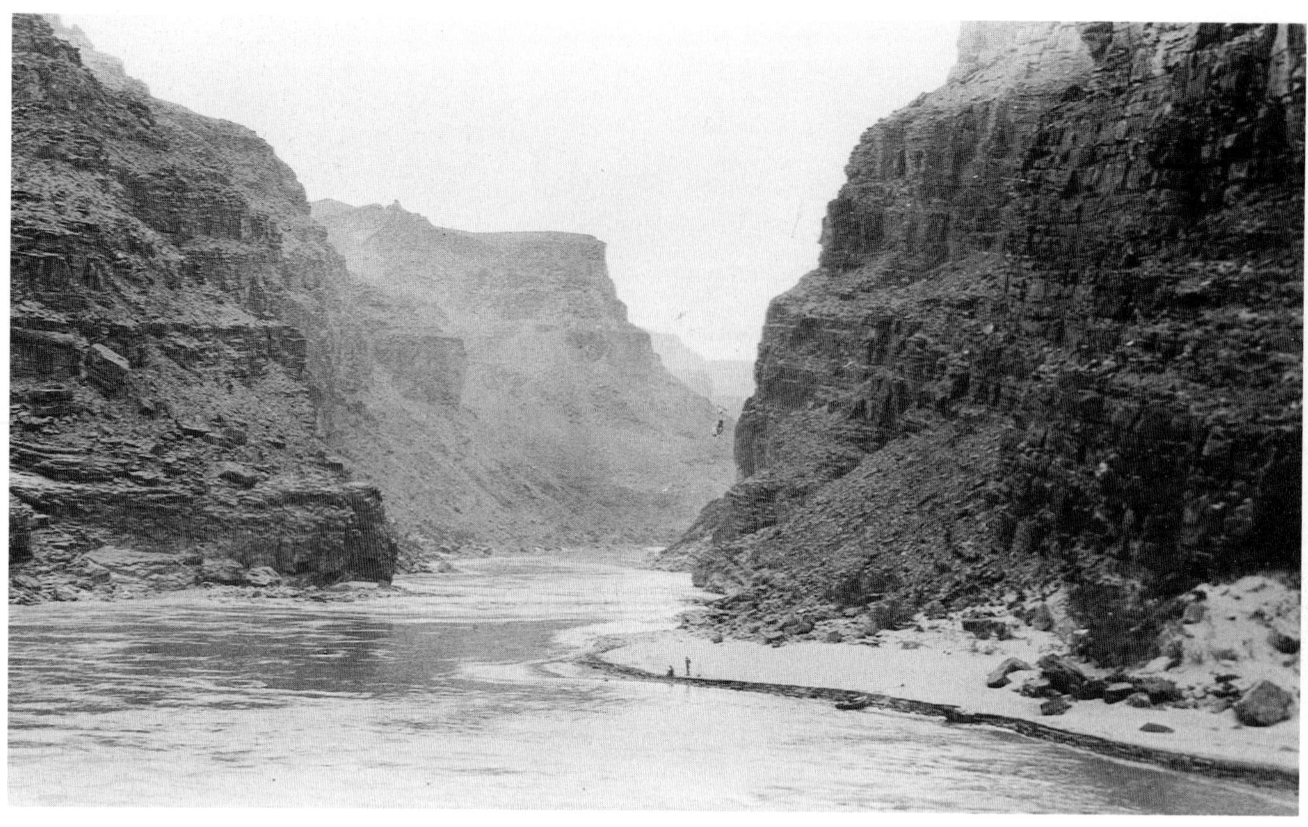

23. Algonkian Gorge above Hance Rapid, mile 75.7, September 13, 1991, 3:00 P.M.

This photo—the original version was taken on the morning of August 16, 1923—looks downriver to the west toward Hance Rapids, just out of sight around the far bend. At this point the Colorado makes a right-angle bend and heads generally to the west. The photo was taken about one hundred feet above the river from a ledge of black, thin-bedded Shinumo Quartzite, part of a Precambrian unit that in earlier stratigraphic nomenclature was referred to as Algonkian (now Proterozoic) in age. These terms refer to upper, or late, Precambrian time, versus the Archean, or Archeozoic, to designate earlier Precambrian time.

Moore drew a cross section at approximately this location, which he referred to as a "projected damsite 1 mile above Hance Rapids." In his notebook he described "considerable bodies of basaltic rock and d[ar]k red ss [sandstone] interbedded lenticular with red sh[ale]. Faults cut these rocks rather frequently."

In the original photo, two members of the survey party are manning an instrument location on the sand spit. Our 1991 shot shows how tamarisks have fixed the sand in place with their roots and extended the spit into the river, thus narrowing the channel and increasing river velocity.

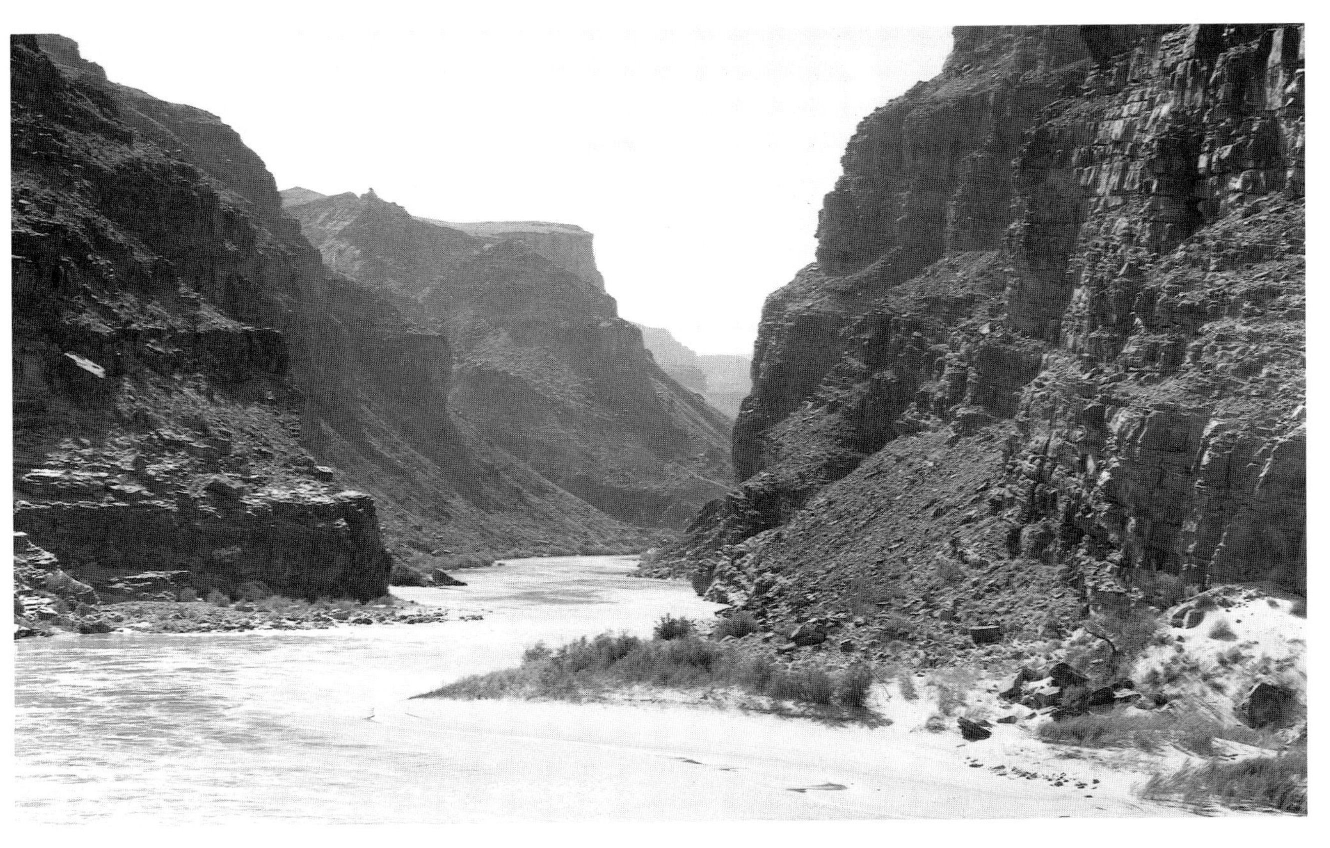

23. Algonkian Gorge above Hance Rapid, mile 75.7, September 13, 1991, 3:00 P.M.

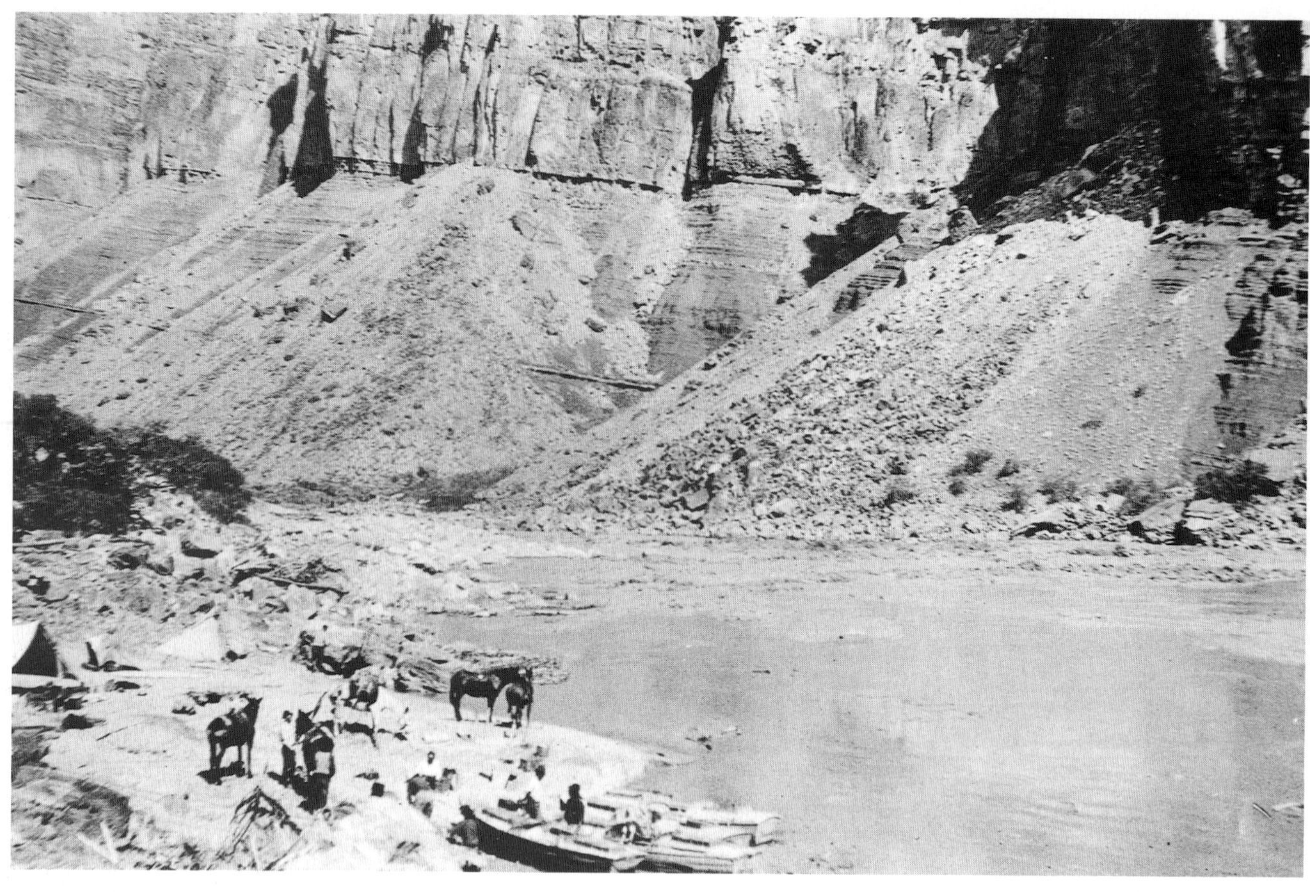

24. Top of Hance Rapid, mile 76.5, September 13, 1991, 5:00 P.M.

This 1923 photo was taken at about noon on August 19 from a sandy bench some fifty feet above the south bank of the river, looking north. Three members of the 1923 crew took a holiday hike from here up Hance Trail to the South Rim, highlighted by attending a "picture show." They returned with a pack train to resupply the expedition; some of the mules stand in the foreground of the picture. The crew portaged supplies around the rapids and then ran their lightened boats through.

Although the higher river level in the 1923 photo obscures many of the rocks now visible in the rapids, it does appear that some have been moved. Both photos show the Shinumo Quartzite cliffs underlain by the slope-forming, Precambrian Hakatai Shale on the opposite canyon wall.

Along with the new growth of tamarisk and sandbar willow in the foreground, older vegetation above the pre-dam high-water line—mostly mesquite—appears to be still in place. The bushes across from the rapids on the lower talus slope look like the same ones in both photographs.

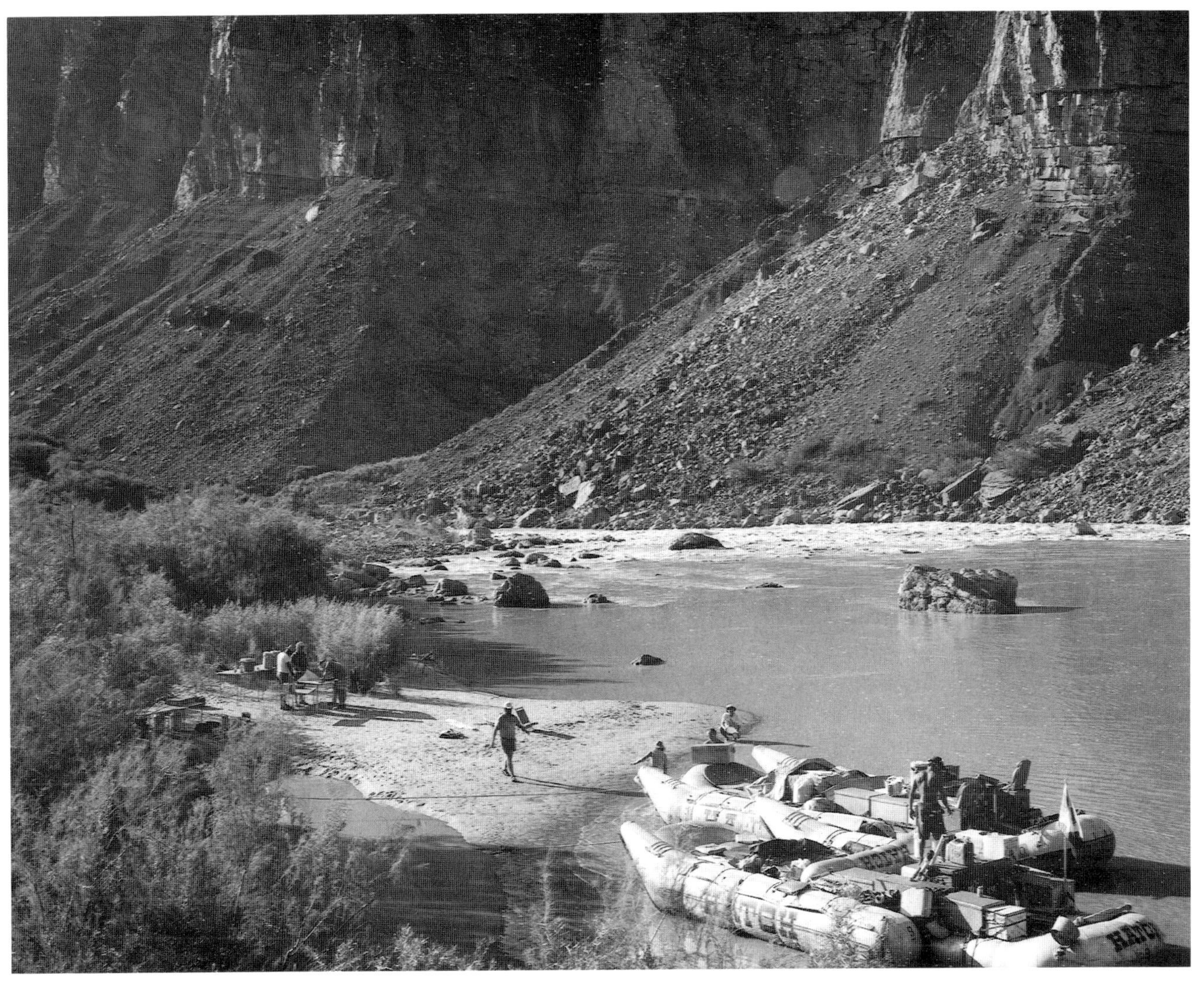

24. Top of Hance Rapid, mile 76.5,
September 13, 1991, 5:00 P.M.

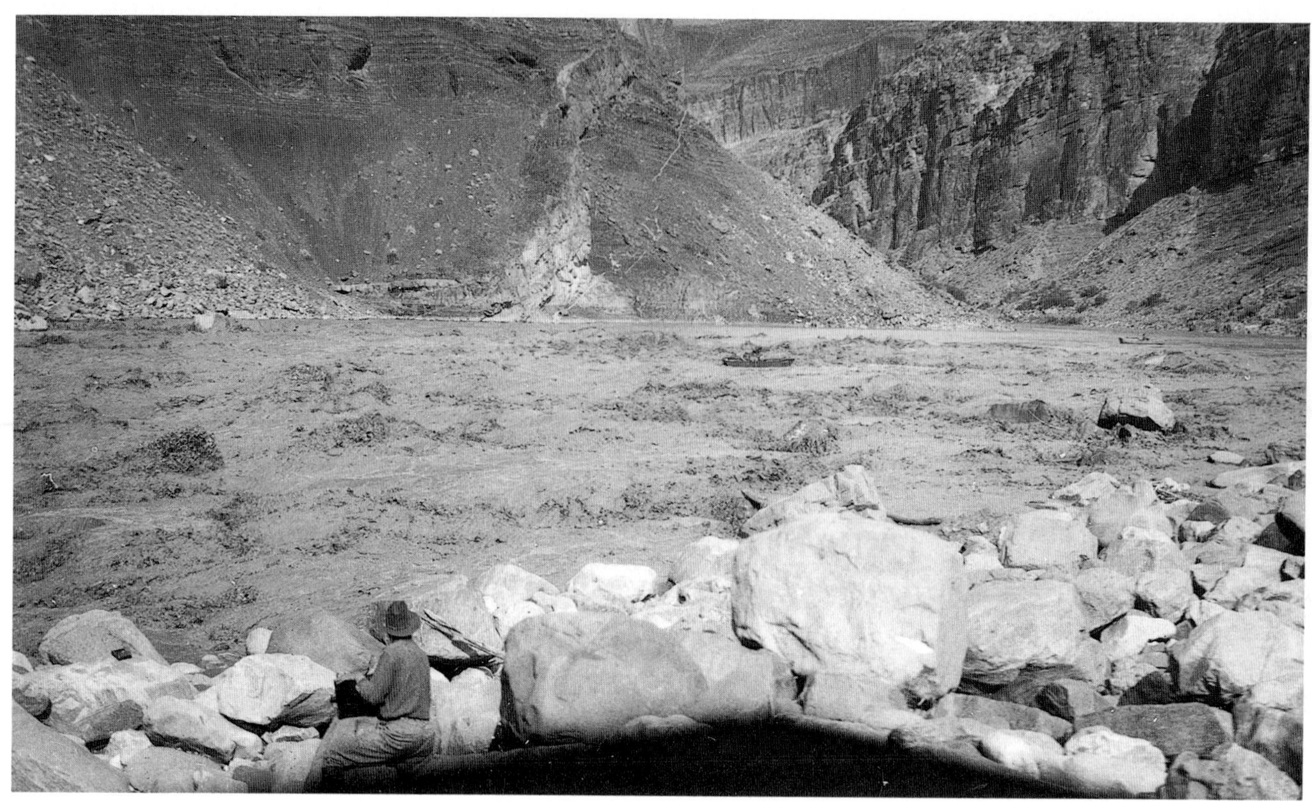

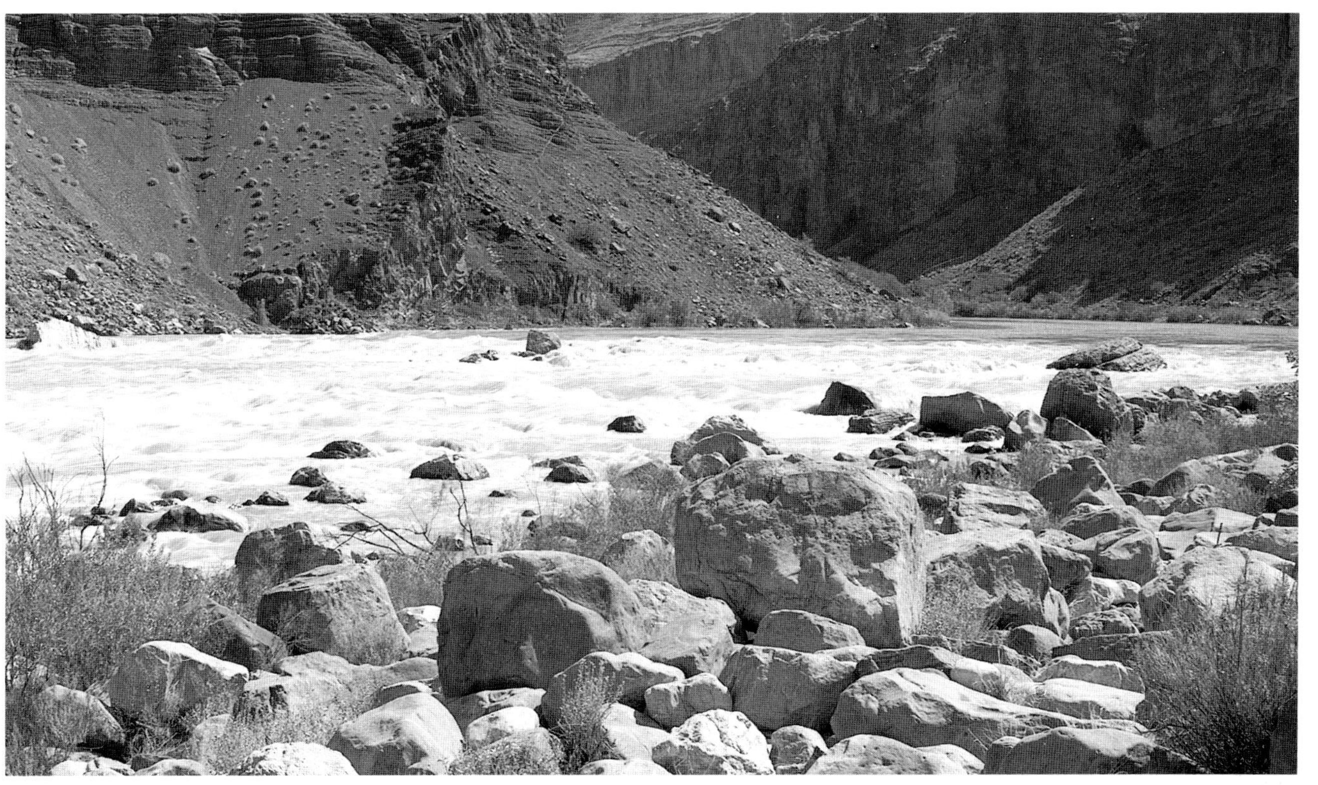

25. Looking upstream at Hance Rapid, mile 76.7, September 14, 1991, 11:00 A.M.

25. Looking upstream at Hance Rapid, mile 76.7, September 14, 1991, 11:00 A.M.

Hance Rapid is an especially long, tough rapid to navigate, rated a nine on the western river runner's scale of difficulty. In his book *Through the Grand Canyon from Wyoming to Mexico,* Ellsworth Kolb described Hance as "a nasty rapid . . . filled with rocks, many of them so concealed in the foam that it was often next to impossible to tell if rocks were there or not, and in which there was little chance of running through without smashing a boat." River runners who work the canyon today have their own stories about the perils of navigating Hance Rapid.

This view, first taken in the early afternoon of August 19, 1923, by Freeman, angles from the south bank, in about the middle of the rapid, back upstream to the east. In the far canyon wall, a thick dike of basaltic igneous rock cuts diagonally upward through the Hakatai Shale; immediately below and parallel to it runs a much thinner stringer dike, barely visible in the older photo.

In the 1923 shot, one of the crew members watches a boat enter the rapid near the center of the photo; another boat can be made out upstream. They are running to the left of a clump of large boulders that is visible in the modern photo midstream at the head of the rapid. Because the left-hand approach is shallow and very rocky, today's larger boats enter the rapid to the right of these boulders and then cut back to the left to avoid large holes farther down the maelstrom. The rocks in the foreground have changed little, and it is possible to match up a number of individual boulders in the two views.

26. Looking across the river at Hance Rapid, mile 76.7, September 14, 1991, 11:00 A.M.

The 1923 version of this photograph was taken in the early afternoon on August 19 from near the same vantage point as that of the previous photo, except that the camera was turned to point across the river, almost due north, to track the progress of the lead boat. The photos depict the Hakatai Shale dotted with boulders that have fallen down from the high wall of Shinumo Quartzite above.

The newer photo shows how much cleaner the river has become since Glen Canyon Dam; notice the whitewater in the rapid as compared with the muddied waters of the older photo. In the earlier picture, the *Marble*—barely discernible as the dark smudge of an oarsman in midstream—narrowly misses a large hole and wave to left of center. These hazards no longer exist, indicating that large boulders have been moved as the rapid matured through time. Except for the arrival of a few salt cedars in the foreground of the newer photo, very little else has changed at this spot since 1923.

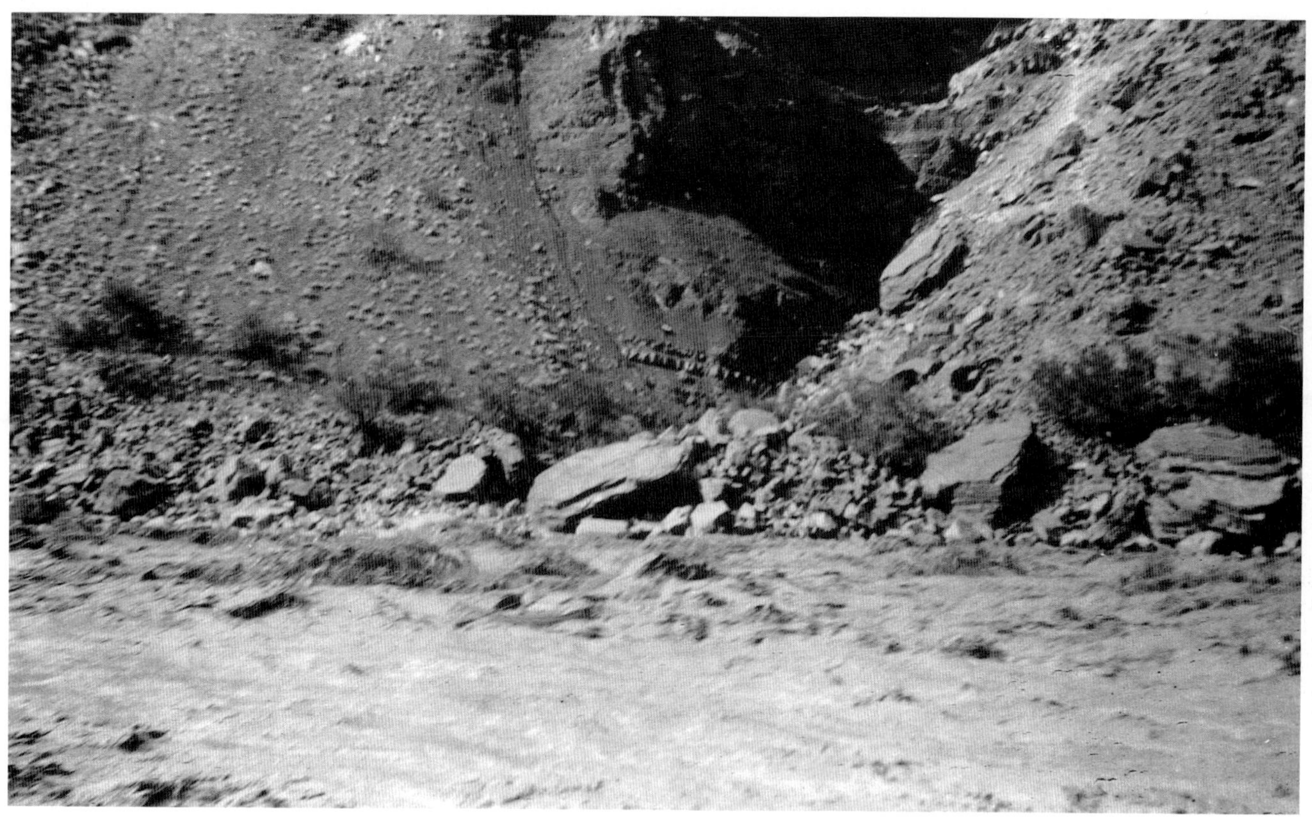

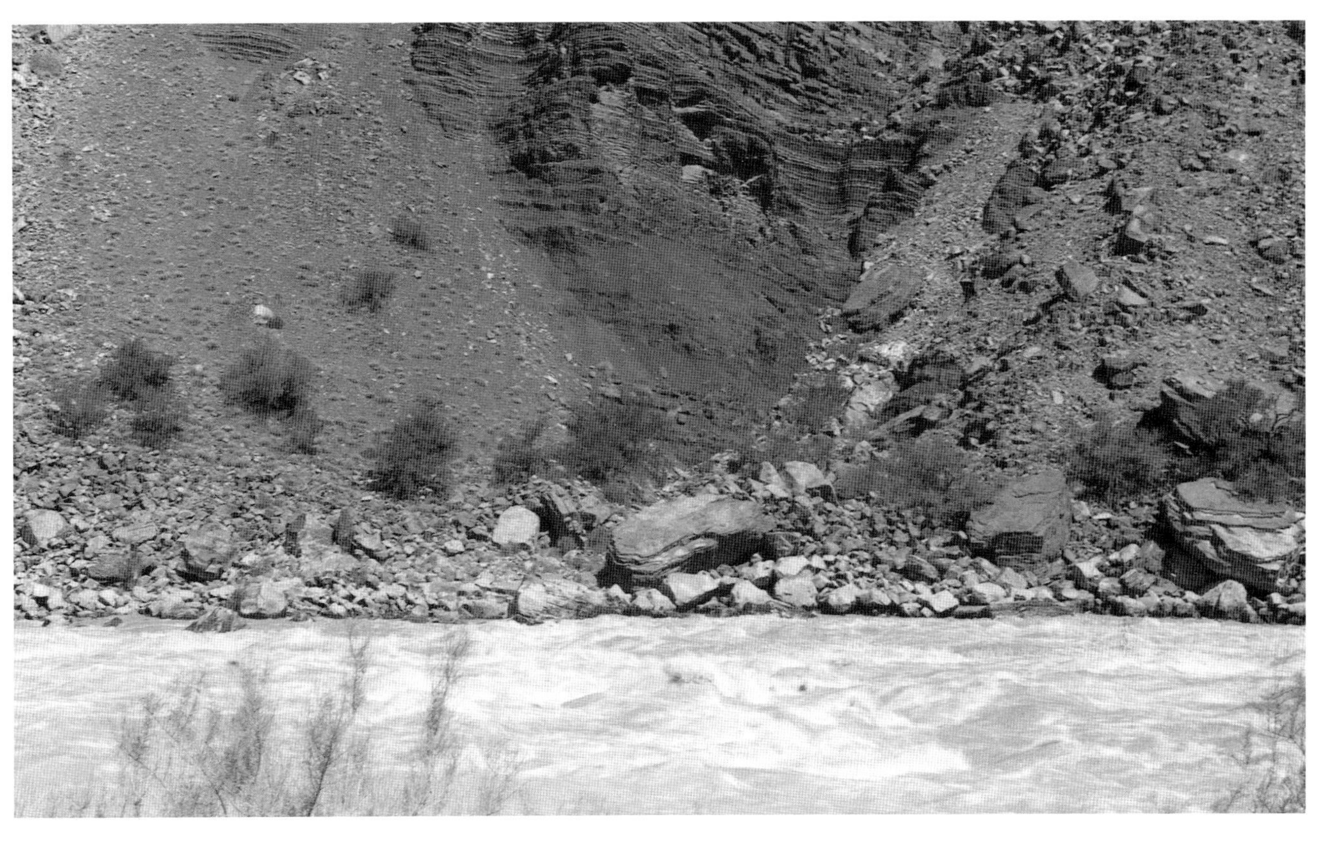

26. Looking across the river at
Hance Rapid, mile 76.7,
September 14, 1991, 11:00 A.M.

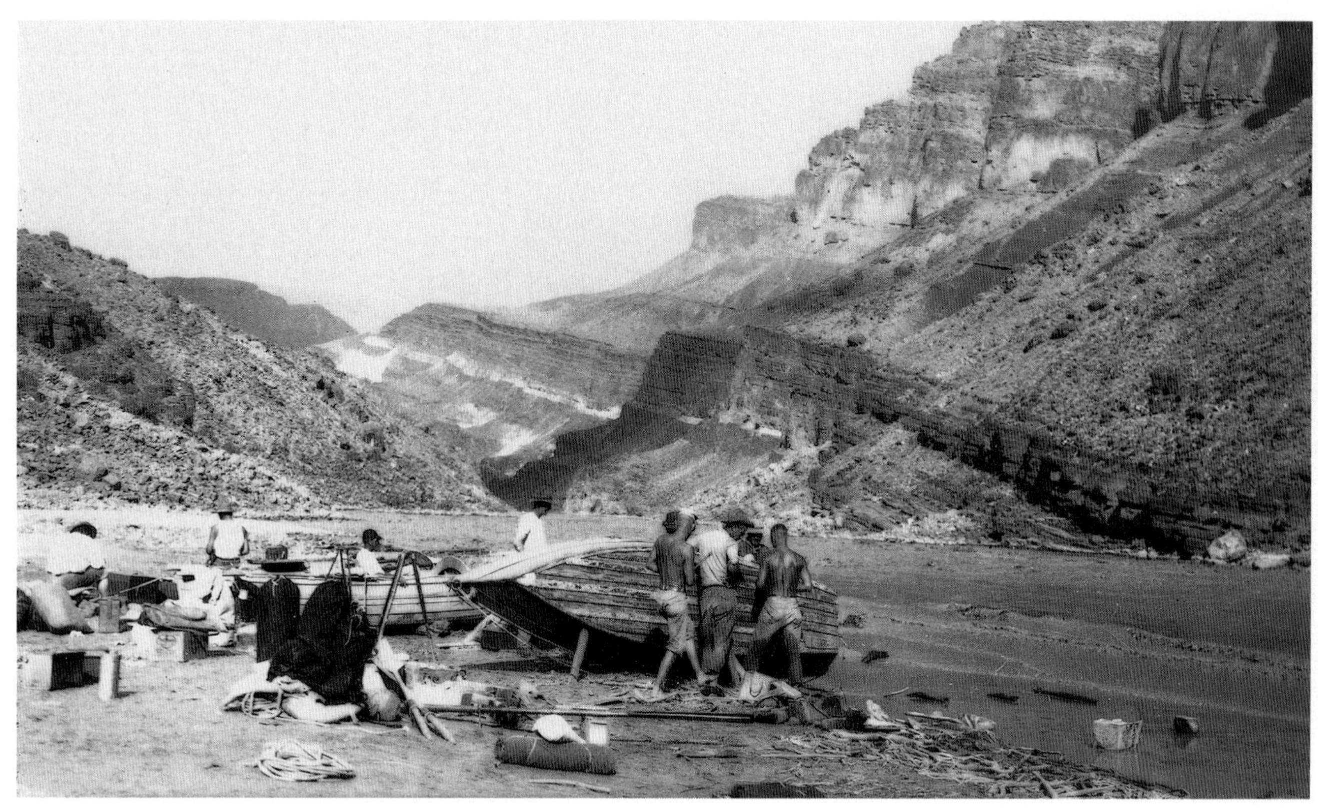

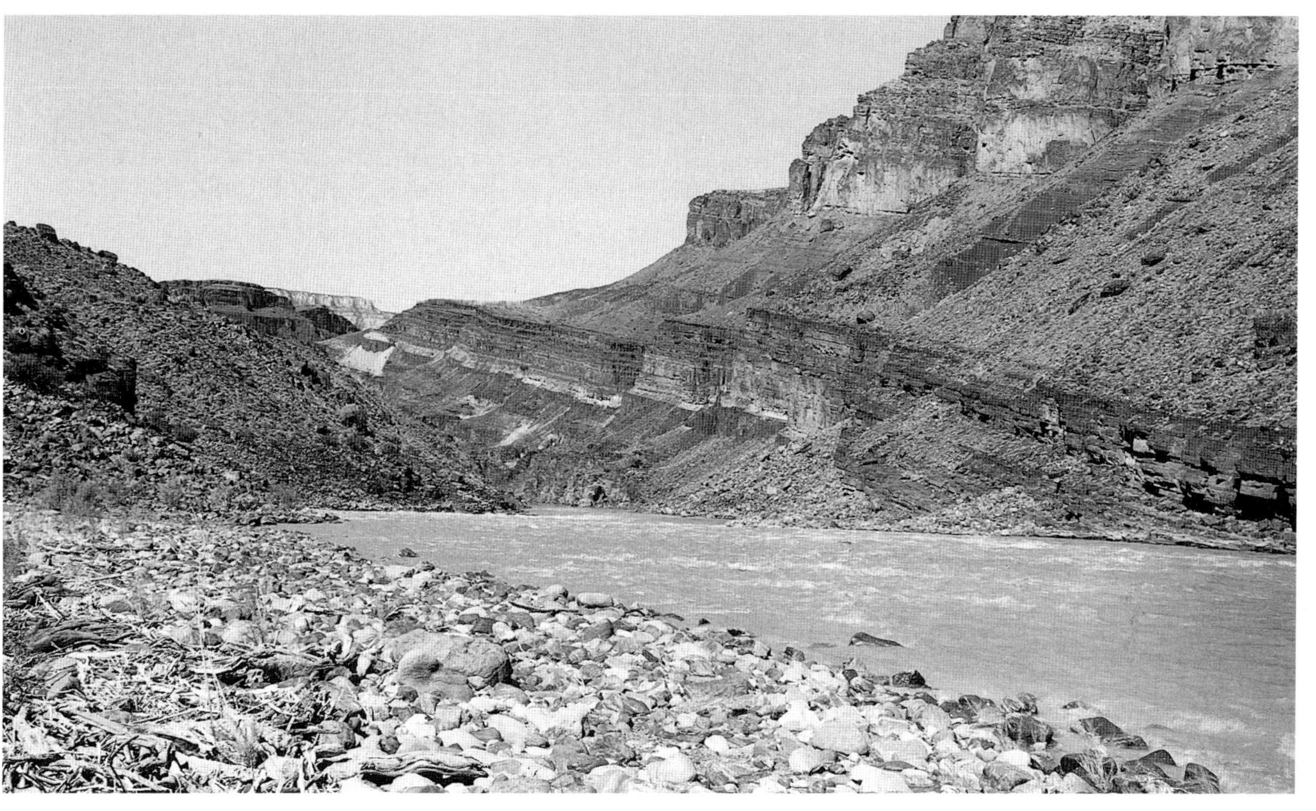

27. Foot of Hance Rapid, mile 76.8,
September 14, 1991, 10:00 A.M.

27. Foot of Hance Rapid, mile 76.8, September 14, 1991, 10:00 A.M.

After running Hance Rapid, the 1923 crew stopped on August 19 to make repairs on the *Boulder.* According to Blake's diary, they also dug several shallow water wells, which accounts for the wood-handled digging tools in the foreground of the older photo.

This view, from the river's edge near the bottom of Hance Rapid, looks downriver, mostly to the west. The Bass Limestone forms the lower cliffs in the far bank. The white band at the base of the cliff is a zone of asbestos formed by contact metamorphism (that is, it baked) when the Bass Limestone was intruded by a diabase sill. This forms the dark slope beneath the white band. In the distance, spoils from an asbestos mine operated by John Hance, the first permanent settler on the South Rim, are visible as a white scar unchanged since 1923.

Sand on the beach in the older photo has now been eroded away; otherwise, recognizable changes are remarkably few.

These photos were taken just above the river's entry into the Granite Gorge, where black crystalline rocks form sheer walls that can still seem overpowering to river travelers. Near here, John Wesley Powell wrote: "August 13, 1869—We are now ready to start on our way down the Great Unknown. Our boats, tied to a common stake, are chafing each other, as they are tossed by the fretful river. . . . We are three-quarters of a mile in the depths of the earth, and the great river shrinks into insignificance, as it dashes its angry waves against the walls and cliffs, that rise to the world above."

28. Grapevine Rapid, mile 81.8, September 14, 1991, 2:00 P.M.

The older view in this pair was taken in the late morning of August 22, 1923, looking generally toward the east up Grapevine Rapid in Granite Gorge. The canyon walls here drop straight to the river's edge, leaving little room to portage. Frederick Dellenbaugh, who accompanied Powell on his second trip down the river in 1872, described running nearby Sockdolager Rapid in his book *The Romance of the Colorado:* "There was a dropping away of all support, a reeling sensation, and we flew down the declivity with the speed of a locomotive. The gorge was chaos."

George Bradley, a boatman on Powell's first expedition, described the night the crew spent on the bank at Grapevine Rapid: "We have but poor accommodations for sleeping tonight. No two except Major and Jack can find space wide enough to make a double bed and if they don't lie still we shall hear something drop and find one of them in the river before morning" (Cooley 1988:157).

The cliffs forming the steep walls of the lower canyon here are on the Vishnu Schist, which dates to nearly two billion years ago. It is capped at the near skyline by vertical, ledgy cliffs of Tapeats Sandstone. The smoothly worn Vishnu Schist made a slick surface from which to work in taking this photo.

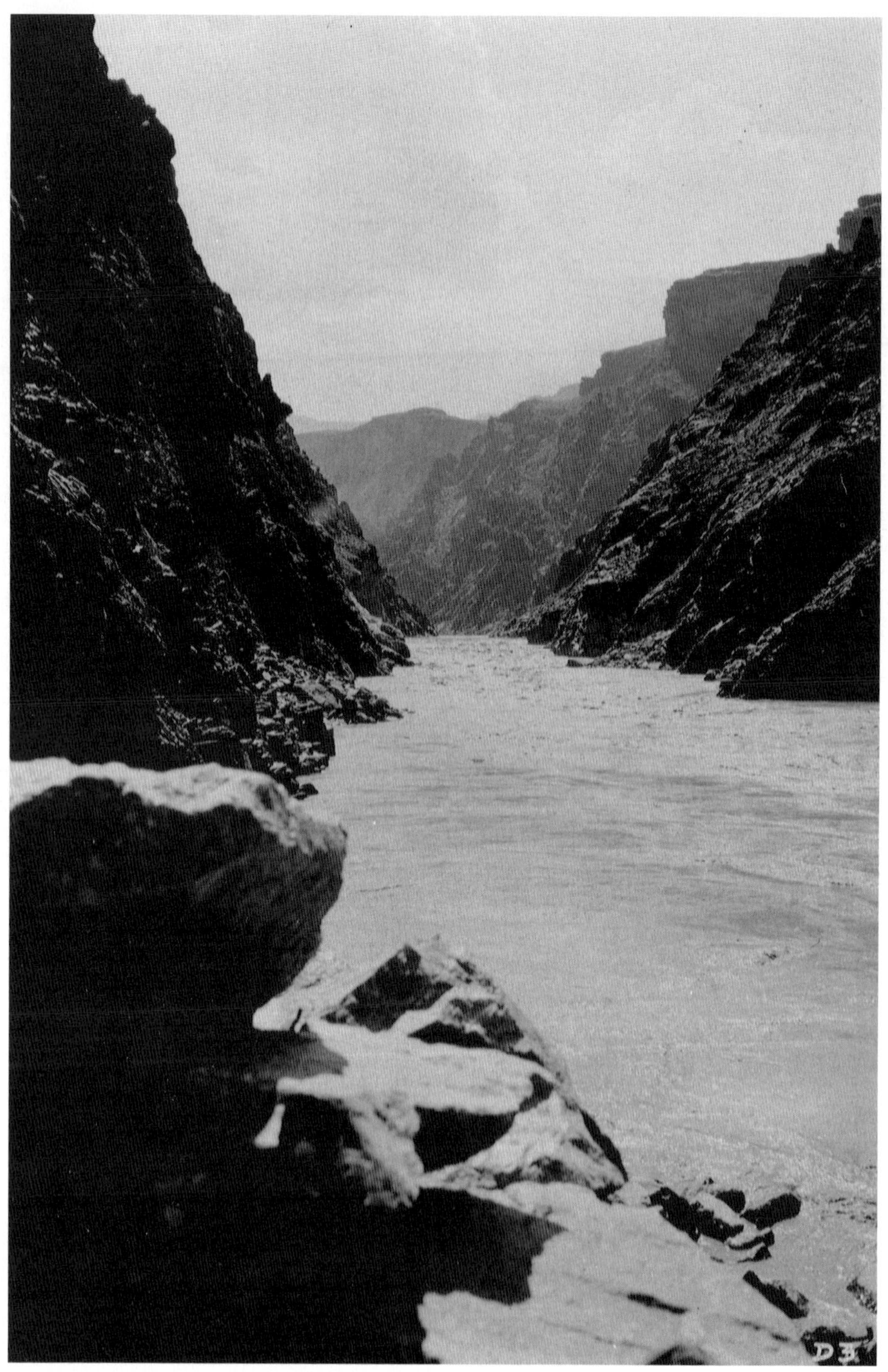
D3

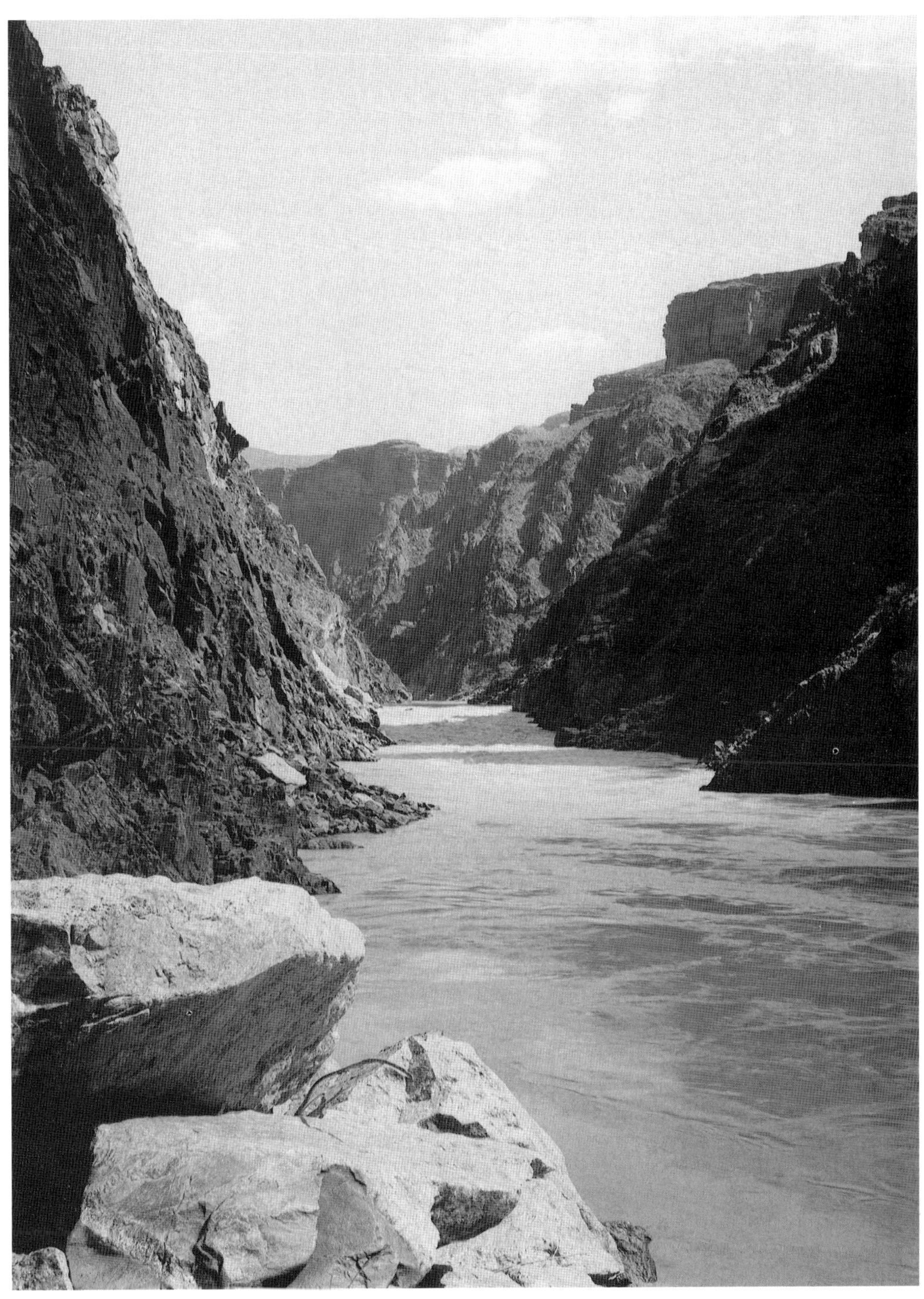

28. Grapevine Rapid, mile 81.8, September 14, 1991, 2:00 P.M.

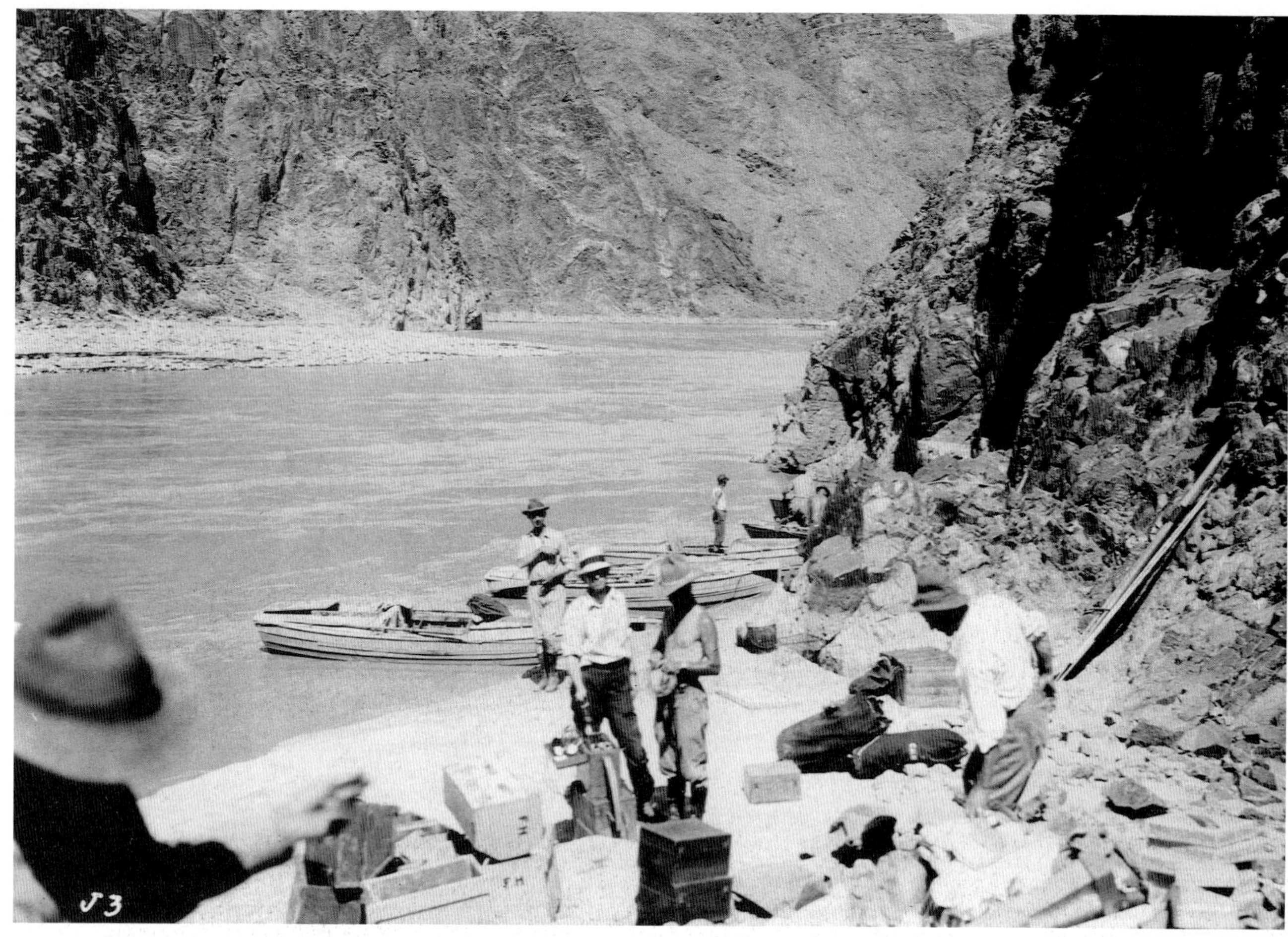
J3

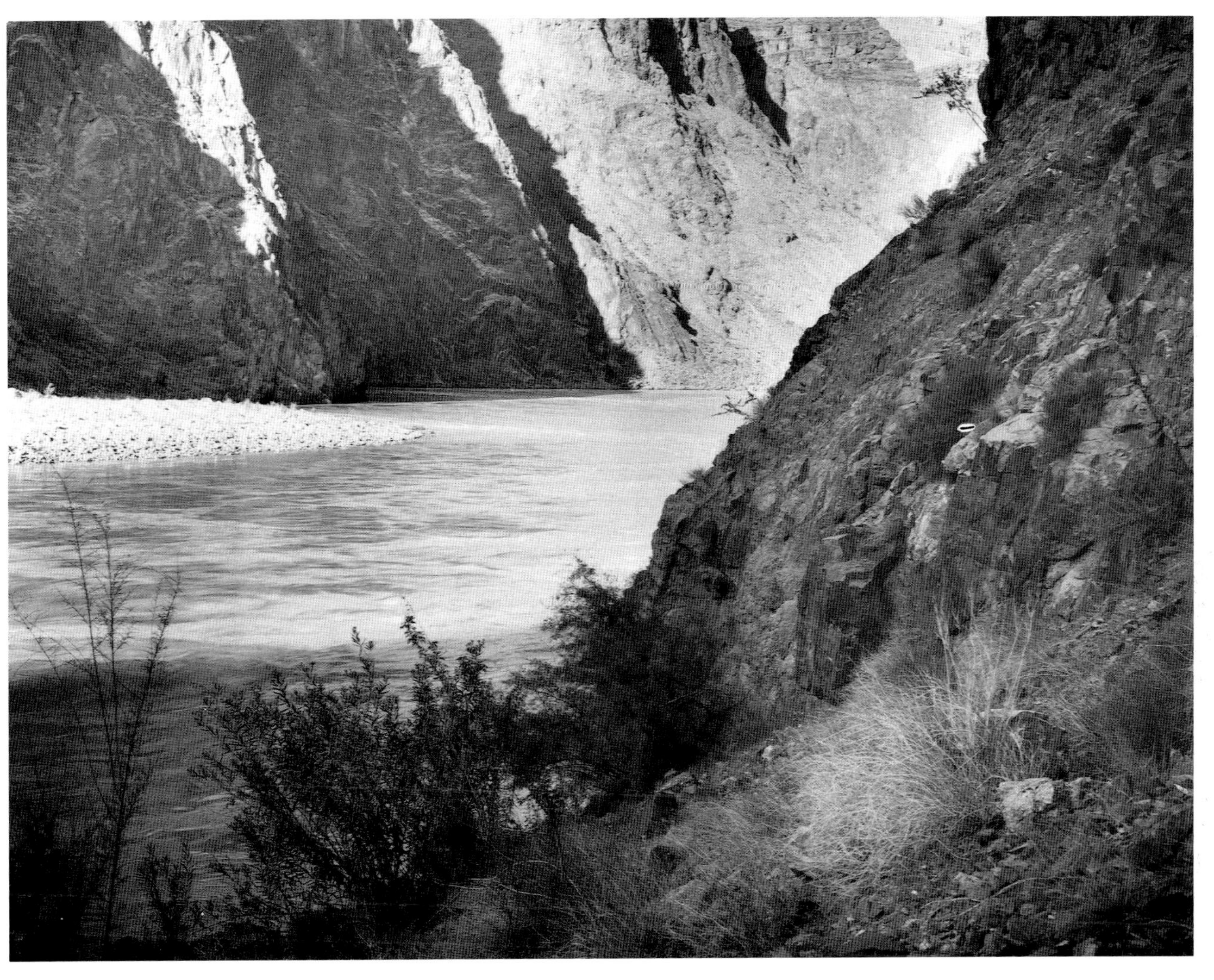

29. Camp at the foot of Bright Angel Trail, mile 88.9, September 14, 1991, 5:00 P.M.

29. Camp at the foot of Bright Angel Trail, mile 88.9, September 14, 1991, 5:00 P.M.

The "official" foot trail to Phantom Ranch in 1923 came down from the South Rim along present-day Bright Angel Trail as far as Indian Gardens and then turned eastward along the Tonto Plateau to the Kaibab Trail, which descends to the river at the upper footbridge. The route of today's lower Bright Angel Trail was only a "secret trail" down Pipe Creek, according to Elwyn Blake's diary. When the new trail was put in by the Civilian Conservation Corps beginning in 1933, man-made talus completely buried the campsite shown in this 1923 photo.

Although we were able to line up general features from the older photo, its camera location now lies about ten feet below the surface of the talus. Everything in the foreground below the black knob of rock (or from just above the standing oars in the lower right of the 1923 photo) is buried by the construction dump.

Looking across the river, the spotty sand veneer on the cobble bank in the older photograph has been washed away, and a sand bank just where the river goes out of view in the center of the older photo is now gone. More than a hundred miles downstream, the effects of Glen Canyon Dam are still apparent.

This view, first taken about midday sometime between August 24 and 27, 1923, faces upstream toward the east from the south bank of the river near the mouth of Pipe Creek. The canyon walls here are dominated by the Vishnu Schist, shot through with dikes of the Zoroaster Granite; a small downfaulted block (graben) of Bass Limestone can be seen just to the left of the notch in the upper right of the photograph.

30. Hermit Rapid, mile 94.9, September 15, 1991, 10:00 A.M.

Birdseye's group reached Hermit Rapid on August 29, 1923, and according to Elwyn Blake was met by a group of "Arizona state and Park officials, who will watch us run Hermit Creek Rapid." The older photograph in this pair, taken about mid-morning on August 30 and looking downriver from the south bank, shows several of those officials, along with Edith Kolb, Emery's daughter, watching a boat that has completed the run. Blake later wrote of his ride through these rapids: "I seemed to be doing all right and relaxed a little. Suddenly I was surprised by a big slopper coming over the side, followed by water from some smaller waves. In all, I took five or six inches of water in the cockpit."

Our photograph, taken only a few feet above the river's edge, shows the usual new vegetation and less muddy river than in the 1923 shot. We could not repeat several old photographs taken near here because willows and salt cedars now completely block the view. In the new photo, boulders and cobbles in the foreground are unrecognizable after readjustment and removal by floodwaters before 1963.

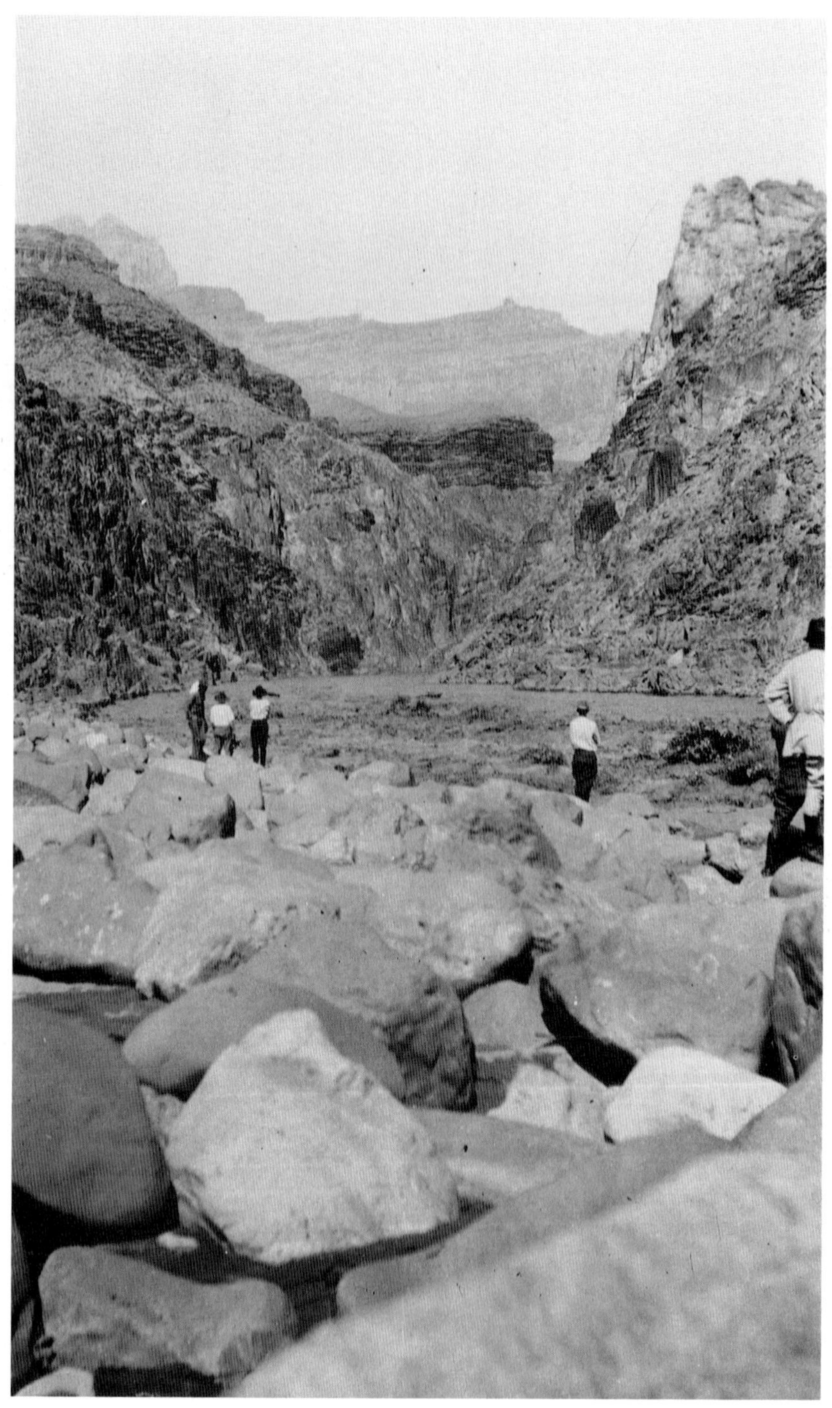

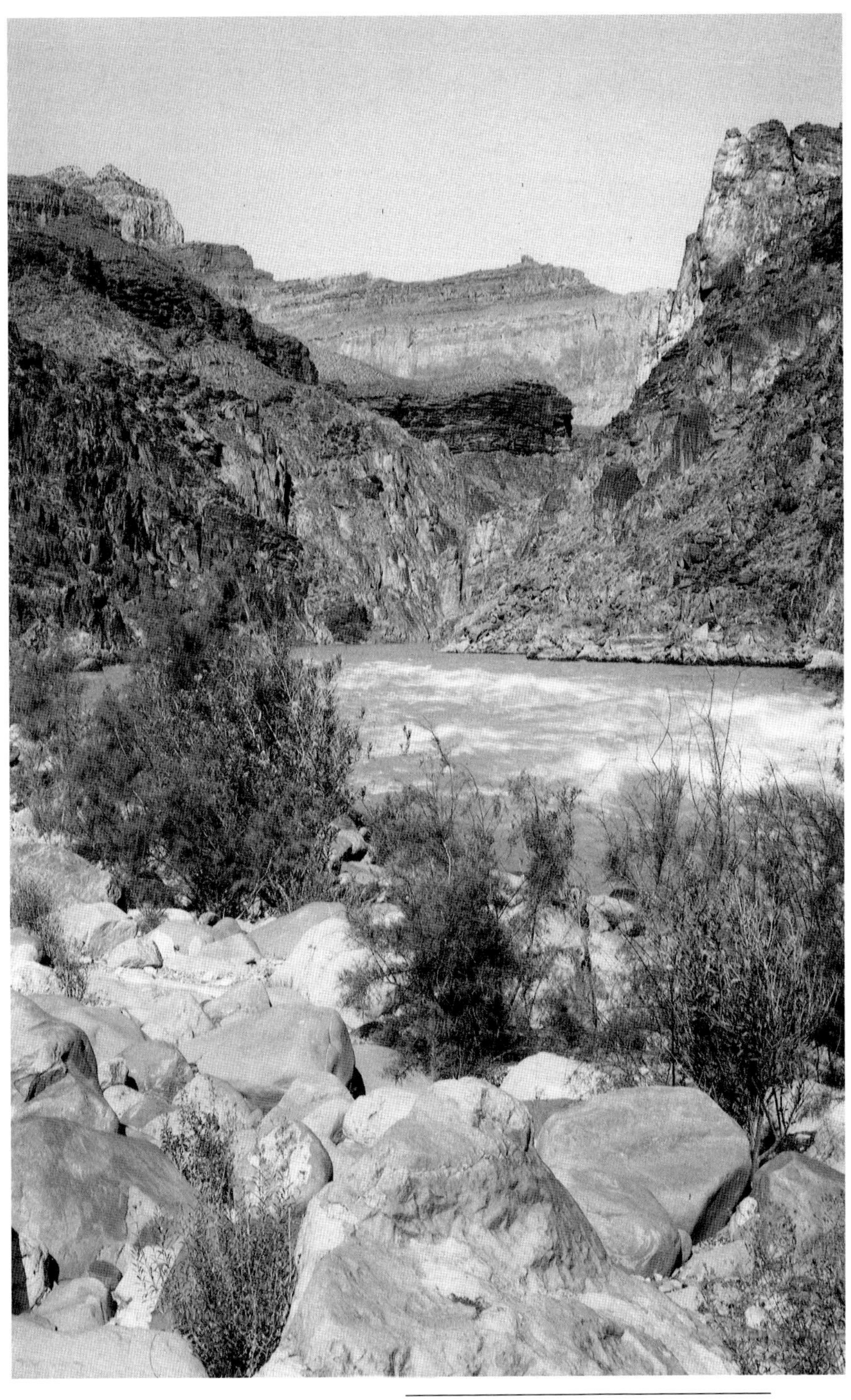

30. Hermit Rapid, mile 94.9,
September 15, 1991,
10:00 A.M.

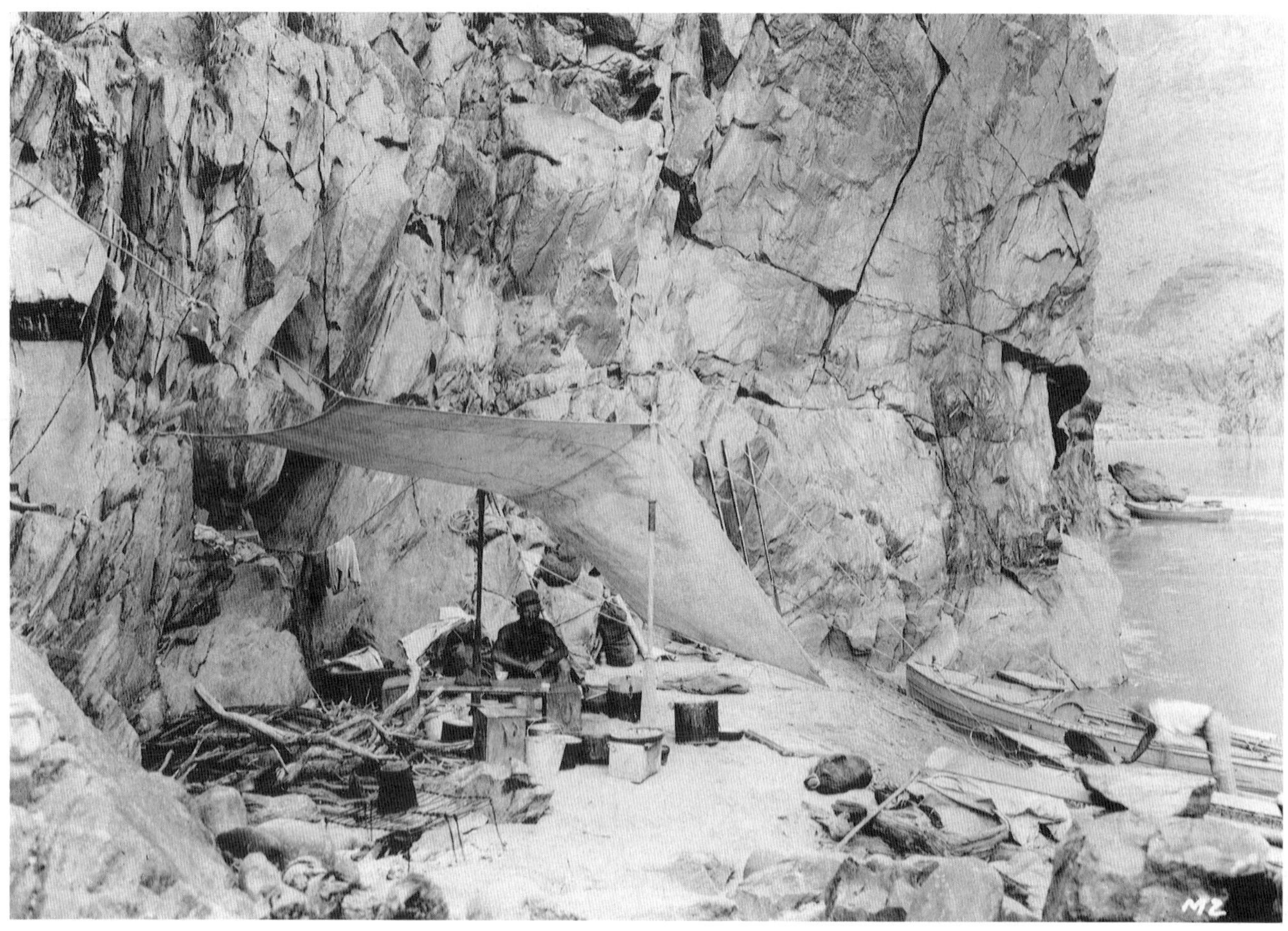

31. Shinumo Creek, mile 108.6, September 15, 1991, 3:00 P.M.

The 1923 party camped on this small sandbank just above the mouth of Shinumo Creek on September 2, and several crew members walked a short way up the creek to see a small waterfall and gather driftwood for fuel. Their camping place is still a popular stopping point for river tourists and receives heavy foot traffic.

This view, originally photographed in mid-afternoon from the north side of the river looking back upstream, or east, again highlights the resistance to change of the Vishnu Schist. Tamarisks and grasses have obtained a recent foothold on the sand in the foreground, which appears somewhat reduced in the modern photo, and several of the larger boulders have been rearranged or are missing.

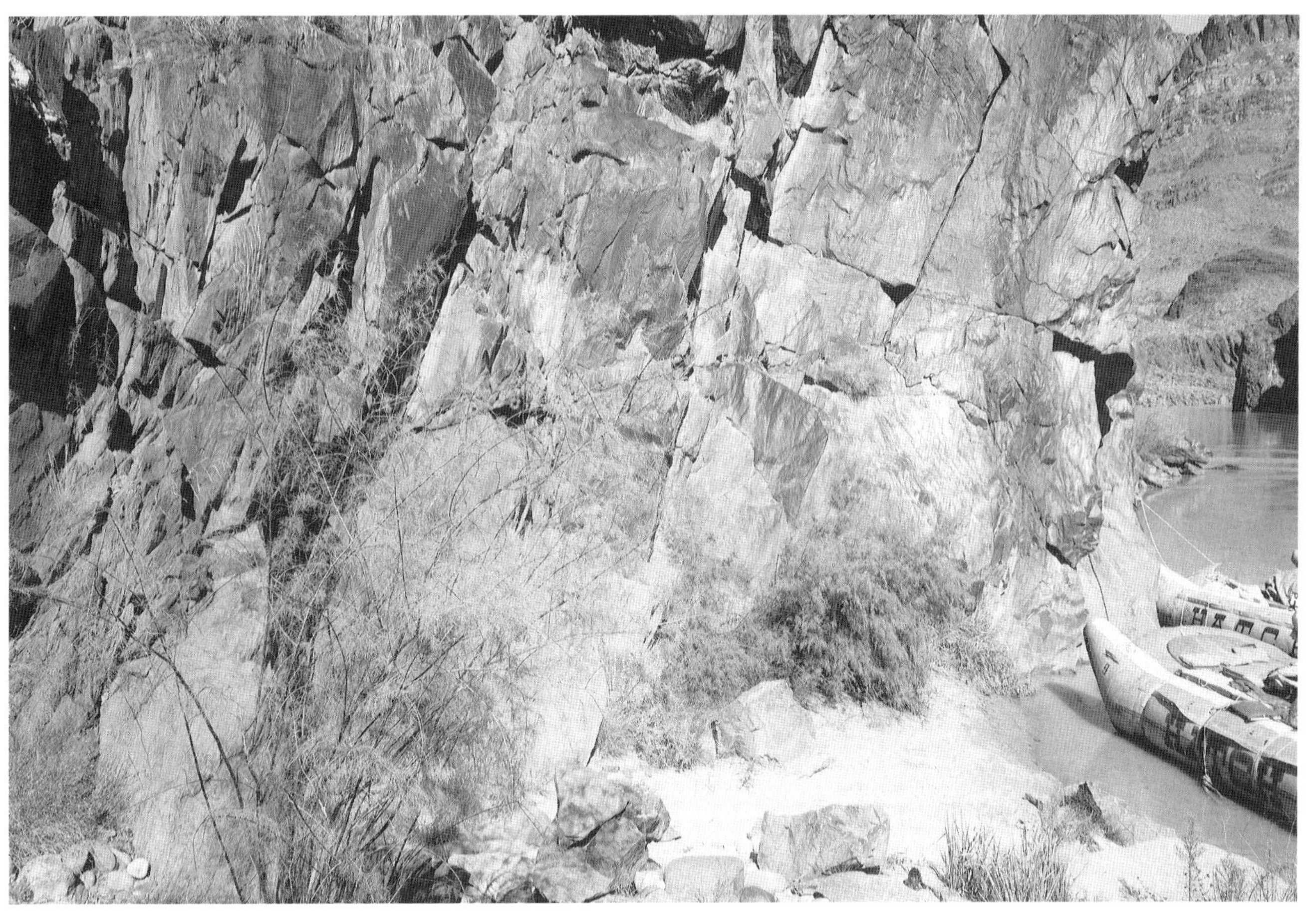

31. Shinumo Creek, mile 108.6,
September 15, 1991, 3:00 P.M.

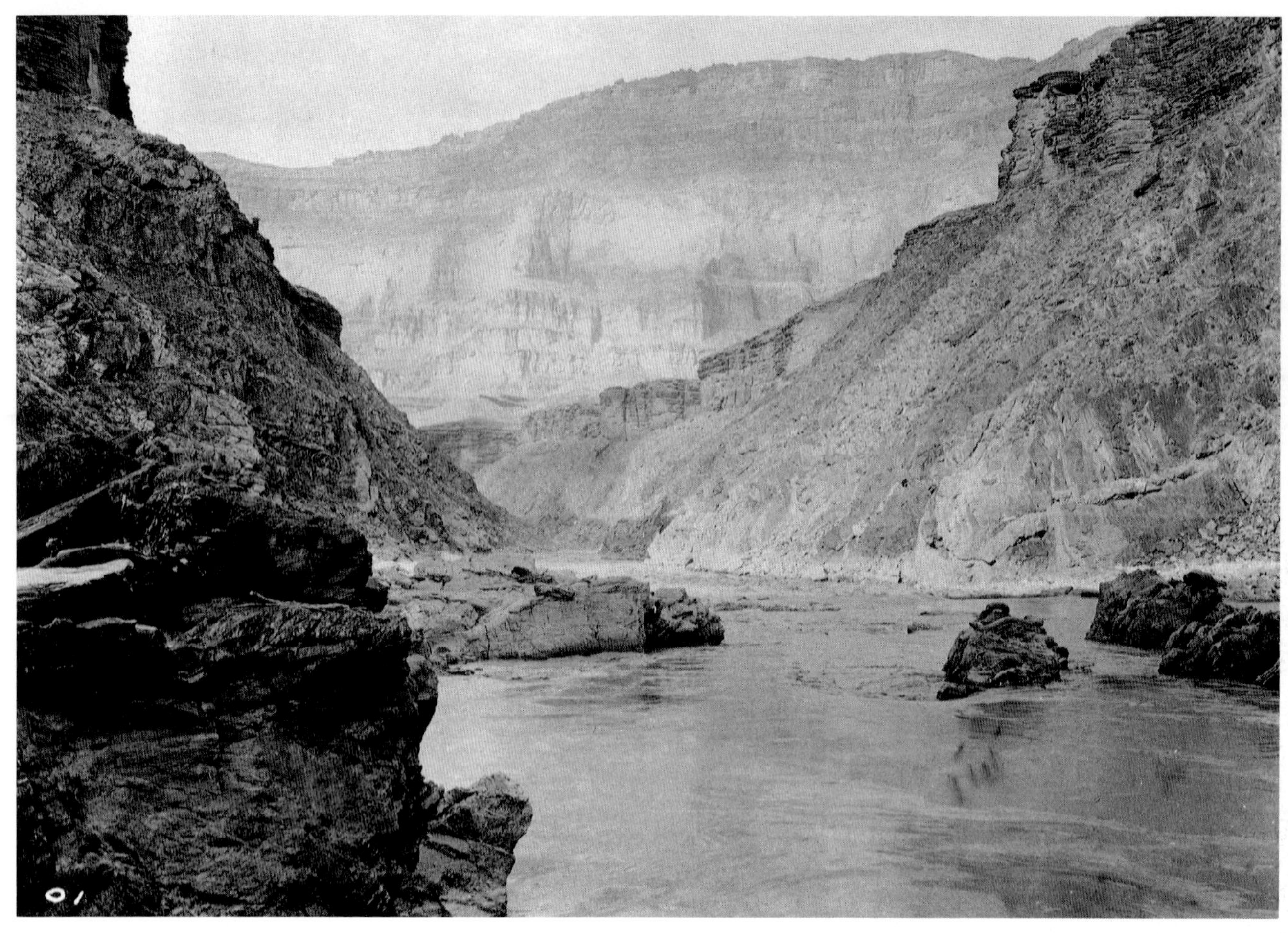

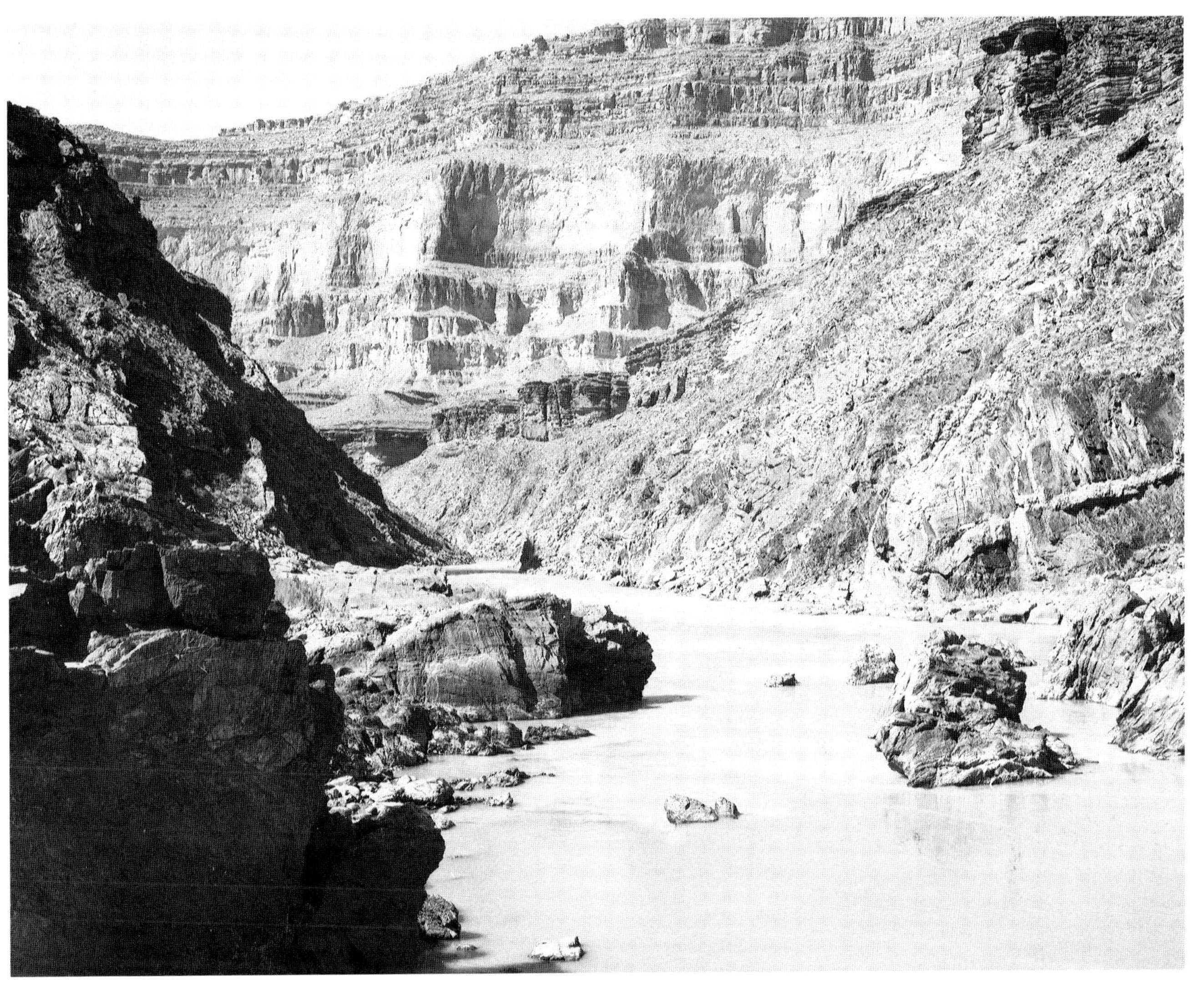

32. Above Waltenberg Rapid,
mile 112.1, September 16, 1991,
10:00 A.M.

32. Above Waltenberg Rapid, mile 112.1, September 16, 1991, 10:00 A.M.

John Walthenberg, who worked with William Bass in establishing Bass Camp and Bass Trail for mining purposes and later with Levi Noble, who published an early geological map of the Shinumo area, lent his name to this rapid—with a slight change in spelling. This 1923 photo, probably by Freeman, was taken just above the rapid in the early morning of September 5 from the south side of the river, looking downstream to the west. Here the river still flows through the ancient Vishnu Schist, and the ledgy cliffs that cap the inner gorge in the middle distance are again exposures of Tapeats Sandstone. The rimrock on the skyline is the Esplanade Sandstone.

For the 1923 photo, the camera itself, without a tripod, was perched on a tall projection of schist about fifteen feet above the river's edge. Repeating the photo required holding our tripod from a position on the outside of the ledge, a foothold that gave way during the operation.

When we took the new photograph, the river was running particularly low because of scant discharge from Glen Canyon Dam—exposing a number of rocks that are below water level in the older photo. This rapid drew little comment from the 1923 crew, probably because the high water made it relatively easy to run.

The precarious vantage point for the next pair of photographs sits on the low cliff on the right-hand side of this photo.

33. High above Waltenberg Rapid, mile 112.3, September 16, 1991, noon.

These pictures, taken at about the same time of day, face upstream to the northeast from high above Waltenberg Rapid on the north side of the river. The older one was shot on September 5, 1923, from a perch of loose rock about a hundred feet above the river—a location typical of Emery Kolb's photos. This cactus-studded spot is so isolated and difficult to reach that we doubted anyone had been back to it since 1923. Like the previous shot, the original photograph was obviously taken with the camera placed directly on a boulder. The vantage point for the previous photo pair was the second rocky point projecting from the right bank into the river at right center. The far butte is capped by the Redwall Limestone, and the high wall above the boats is Vishnu Schist capped by Tapeats Sandstone.

In 1923, the beach here included two large, sandy stretches on either side of the parked boats, instead of the one that is apparent where our outrigged pontoon is moored. The foreground of the 1923 photo also shows a beach and a talus slope less covered by blown sand than they are today.

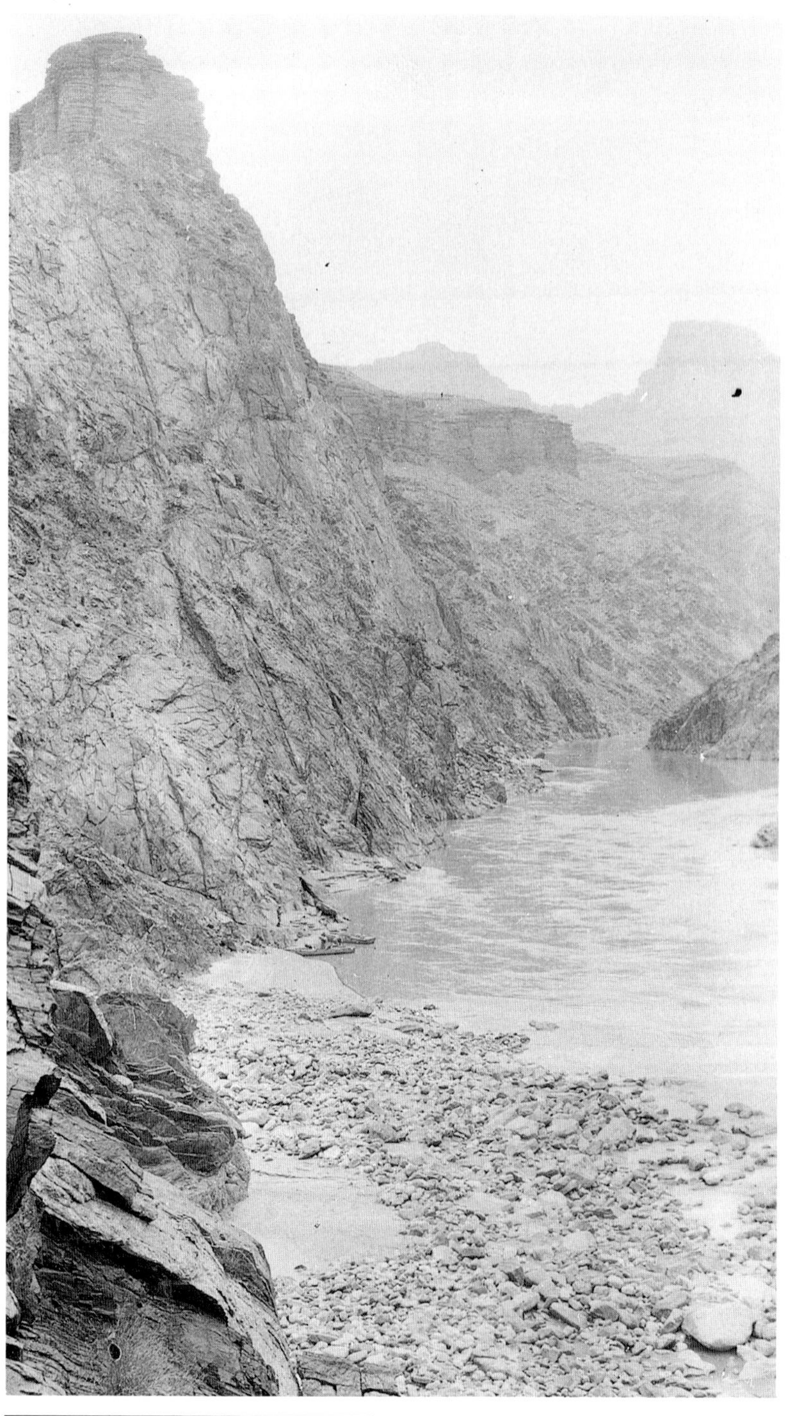

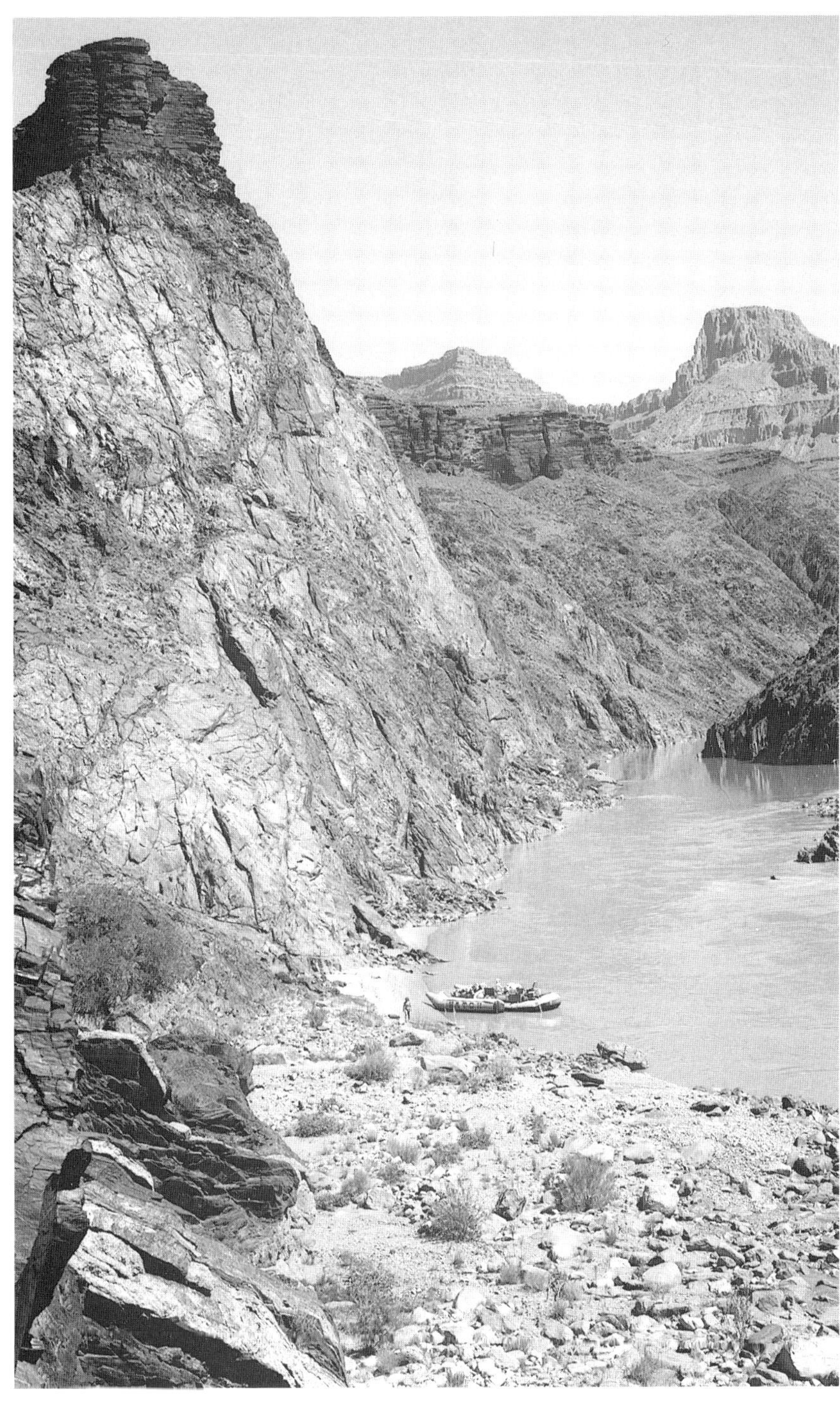

33. High above Waltenberg Rapid, mile 112.3, September 16, 1991, noon.

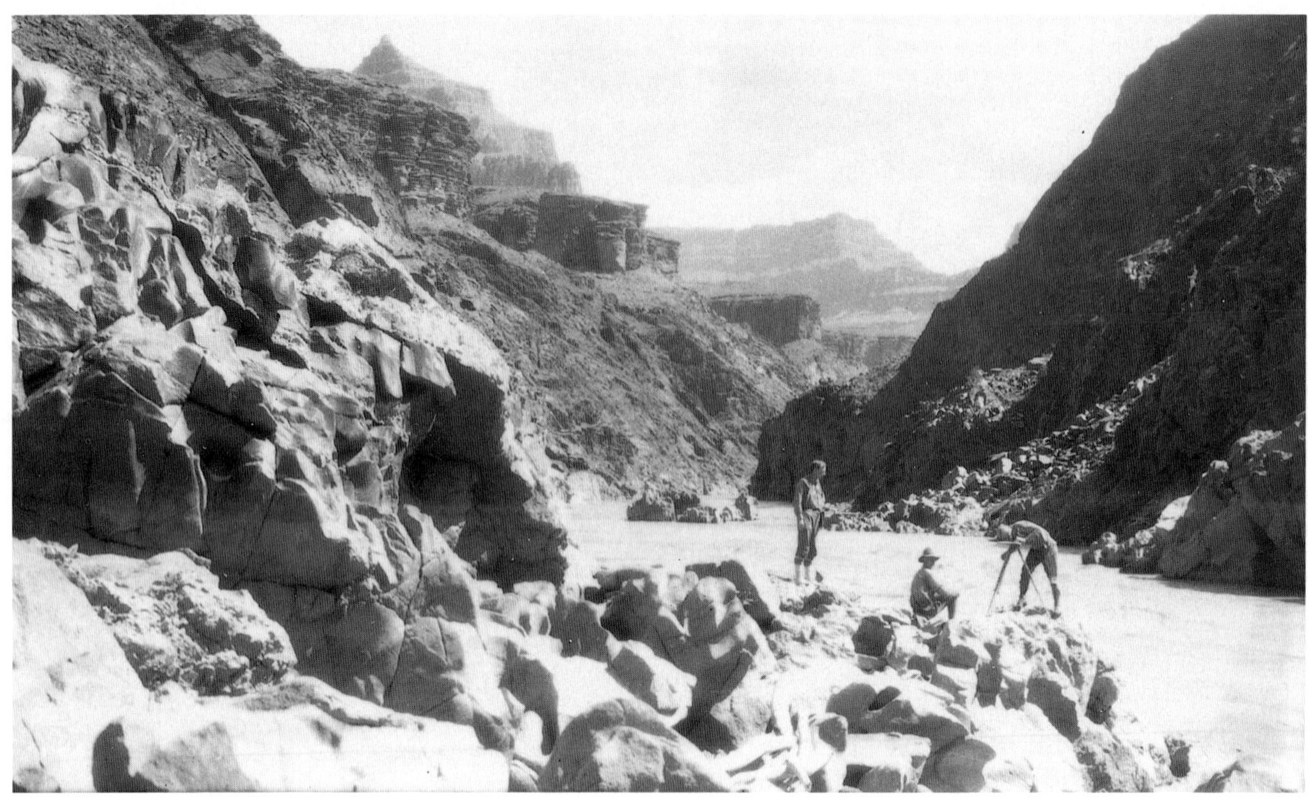

34. Below Waltenberg Rapid, mile 113.1, September 16, 1991, 3:00 P.M.

In the 1923 photo of this pair, members of the survey crew work with plane table and alidade on the north bank of the river in the late morning of September 5. The photo faces back upstream to the east.

Moore looked at a dam site near this location and described the Vishnu Schist: "This rock is very hard and massive, weathering in steep walls which are sheer in places. Joints traverse the gneiss irregularly, defining blocks of various sizes and shapes. The rock of the walls is not shattered and will unquestionably afford highly satisfactory material for dam foundations, abutments and for tunnels. Near water level the rock is finely polished and is irregularly carved in places into potholes etc."

The joints and polish of the rocks are more obvious in the new photo than in the old, as are pods and lenses of light-colored, intrusive pegmatite. Because the crystalline Precambrian layers here defy erosion, they made it possible for us to focus on small details in the rock, using them to line up our repeat photo precisely.

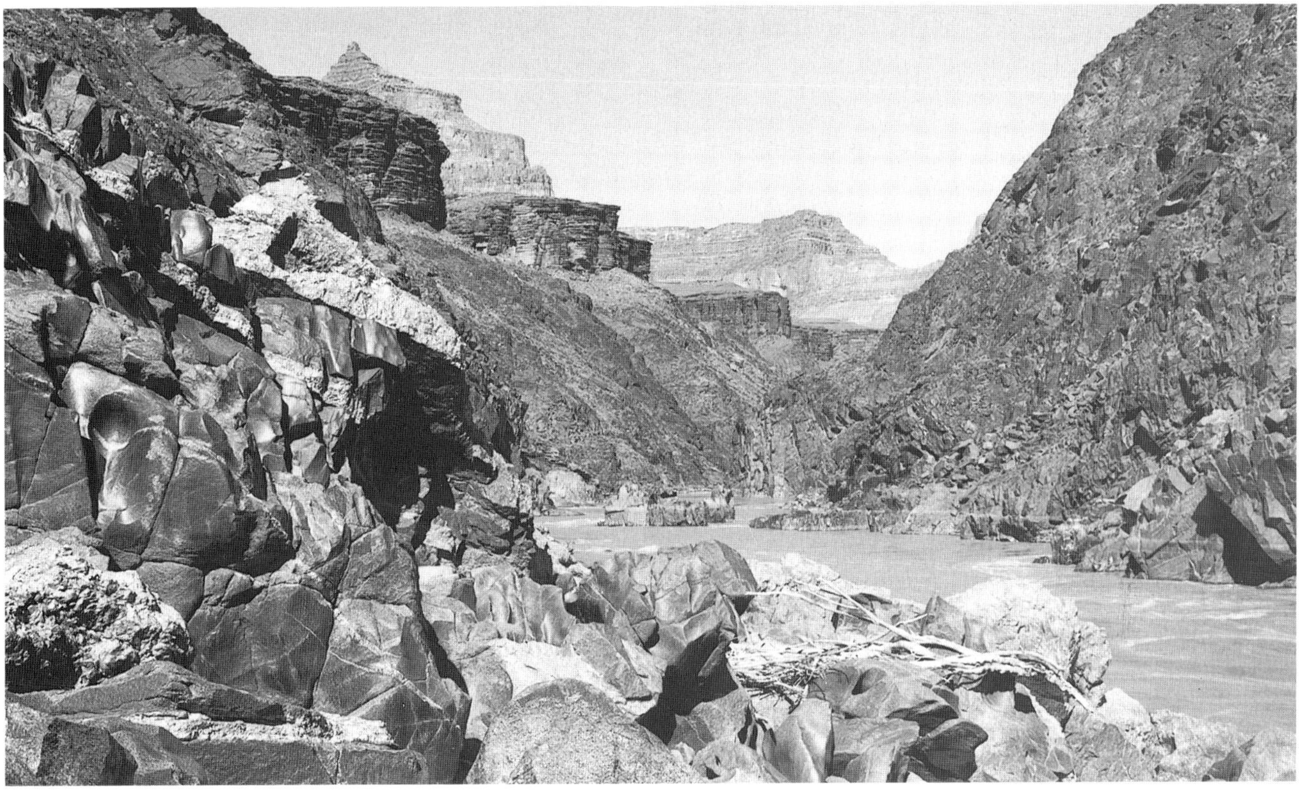

34. Below Waltenberg Rapid,
mile 113.1, September 16, 1991,
3:00 P.M.

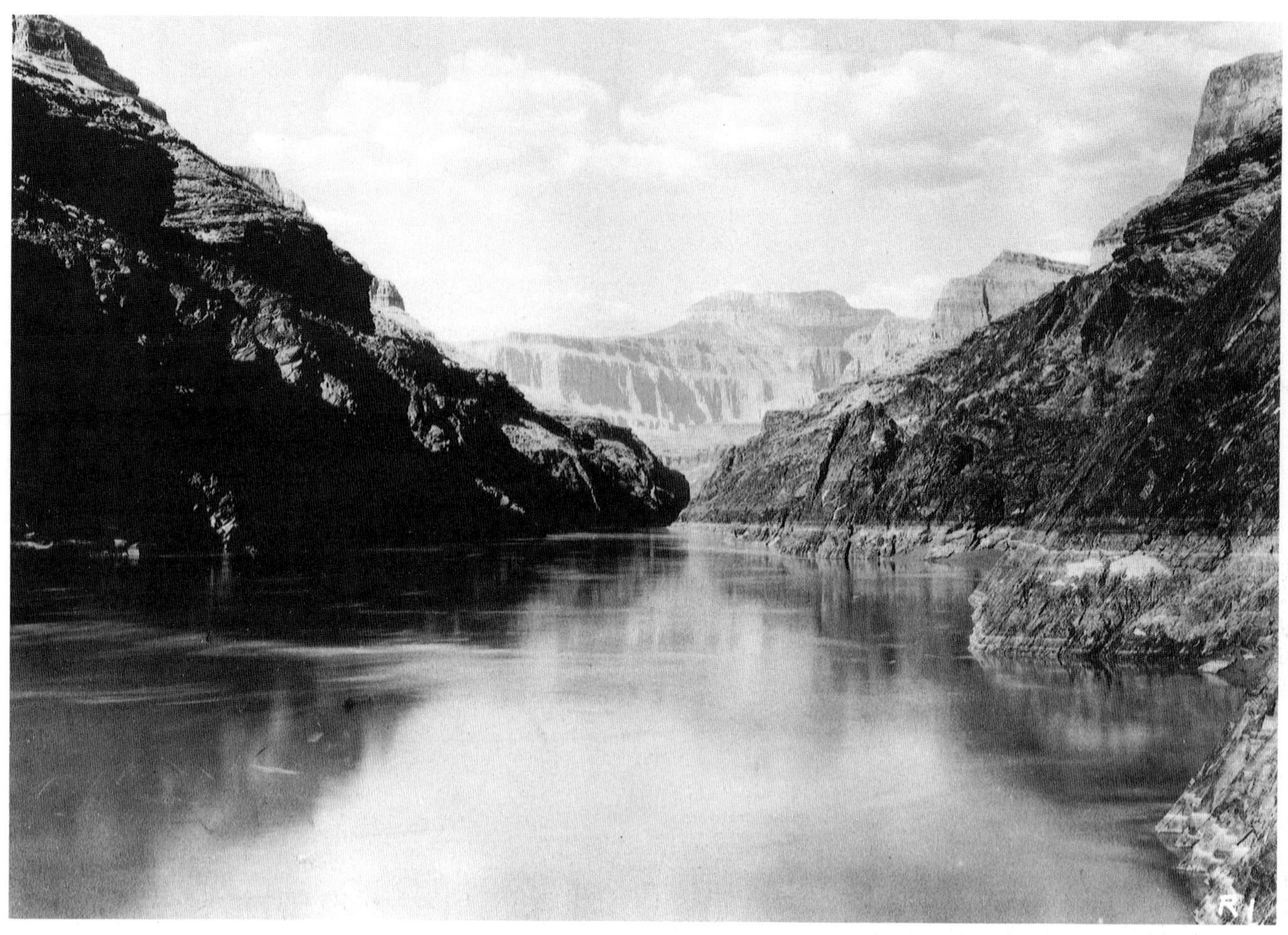

35. Below Fossil Rapid, mile 127.0, September 17, 1991, 9:00 A.M.

These two photographs illustrate a phenomenon called a "bathtub ring" when it is seen on rock walls along dammed reservoirs such as Lake Mead or Lake Powell. The riverbank on the right side of the older photo is marked by lighter-colored rocks below the high-water line; the discoloration comes from suspended clay particles that adhere tenaciously to rock surfaces during times of standing high water. In this location, the discoloration stems from natural rather than human damming, as waters are stilled behind Specter Rapid. The old bathtub ring appears to have washed away by 1991, only to have been replaced at a lower level on the left bank in the new photo.

This view was originally photographed in the early afternoon of September 6, 1923, looking downriver from the east (right) bank at the entrance to middle Granite Gorge. Because of shadows, the two photographs are difficult to compare.

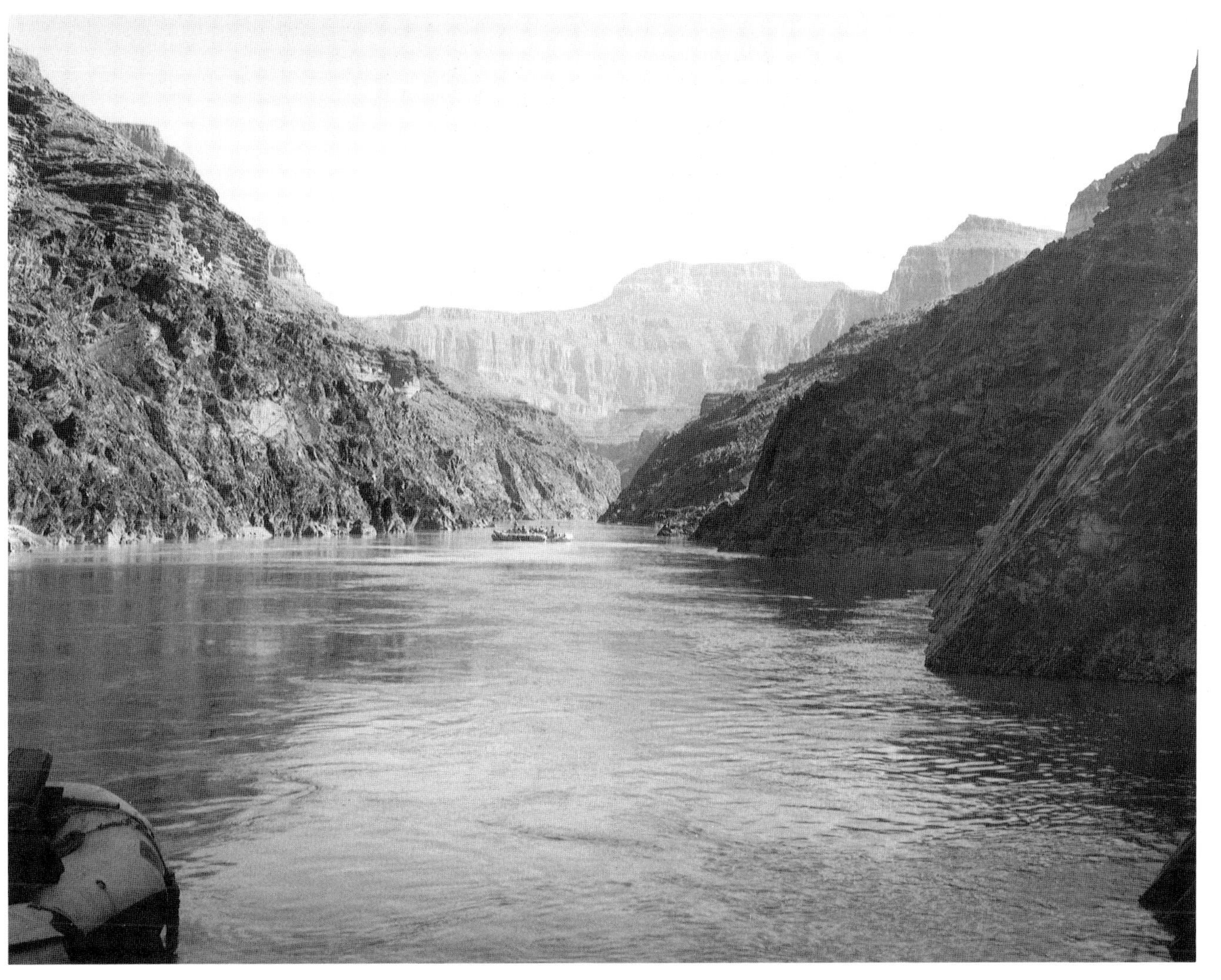

35. Below Fossil Rapid, mile 127.0,
September 17, 1991, 9:00 A.M.

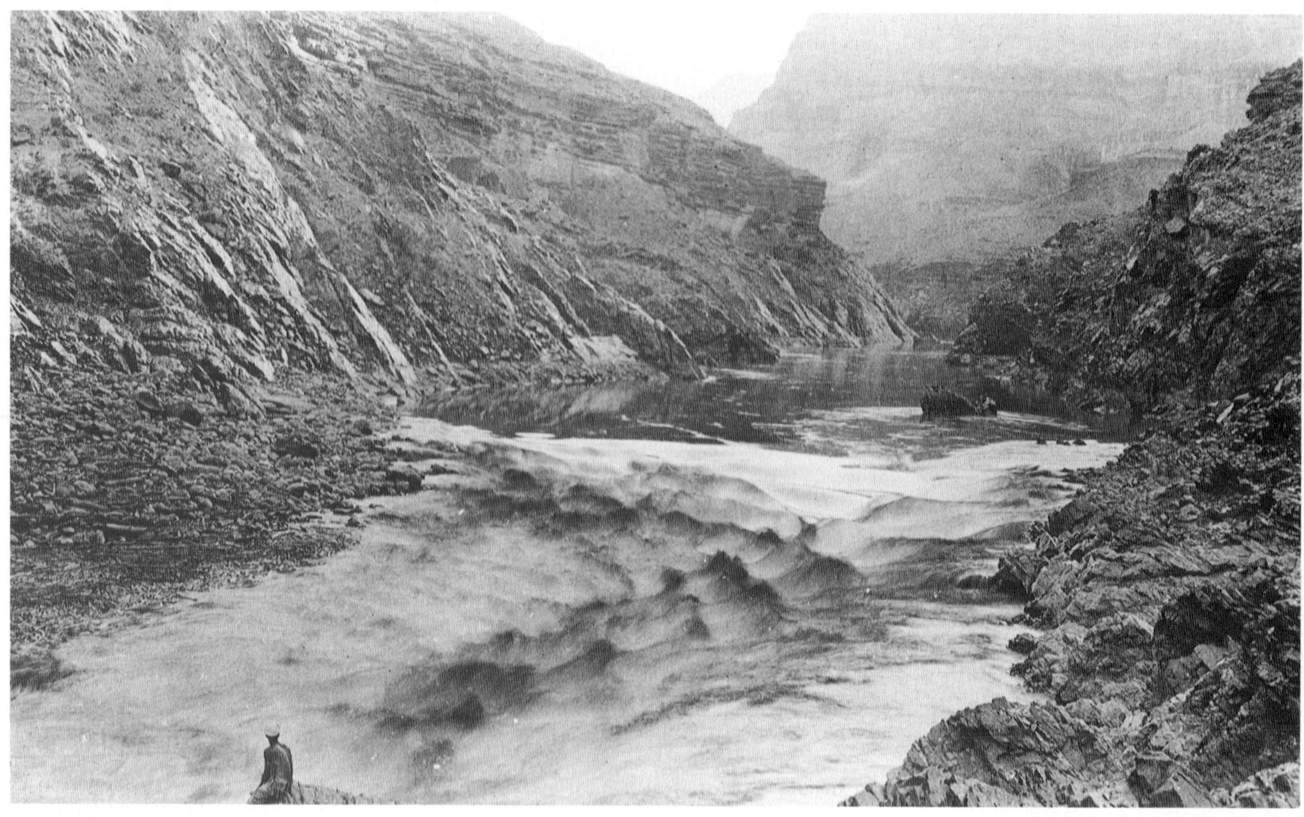

36. 128-Mile Rapid, mile 128.5, September 17, 1991, 11:00 A.M.

After the 1923 crew ran this rapid at a point where the river turns to the northeast, boatman Blake gave the following account: "I profited by having watched the others make their runs. I missed the high, choppy waves entirely by cutting through the left side of the V prow first. I took only one small splash."

The vantage point for this view sits about seventy-five feet above the river on a perch of Vishnu Schist. LaRue took the original shot from the northwest bank, looking back upstream, in mid-afternoon on September 6, 1923. A crew member watches the river in the old photo; Tom McClain provides the 1991 counterpart.

Water level was about the same when these two photos were taken, judging from the row of three rocks that rise just above water level to the far right at the head of the rapid's tongue. In the older photo, substantial driftwood has collected in the back-eddy below the rapid; in the new one, the driftwood is gone and less of the cobble-covered beach is visible. The foliation planes of the Vishnu Schist on the far wall, which shine in the older photo, are obscured in the recent shot, possibly because of additional talus and vegetation.

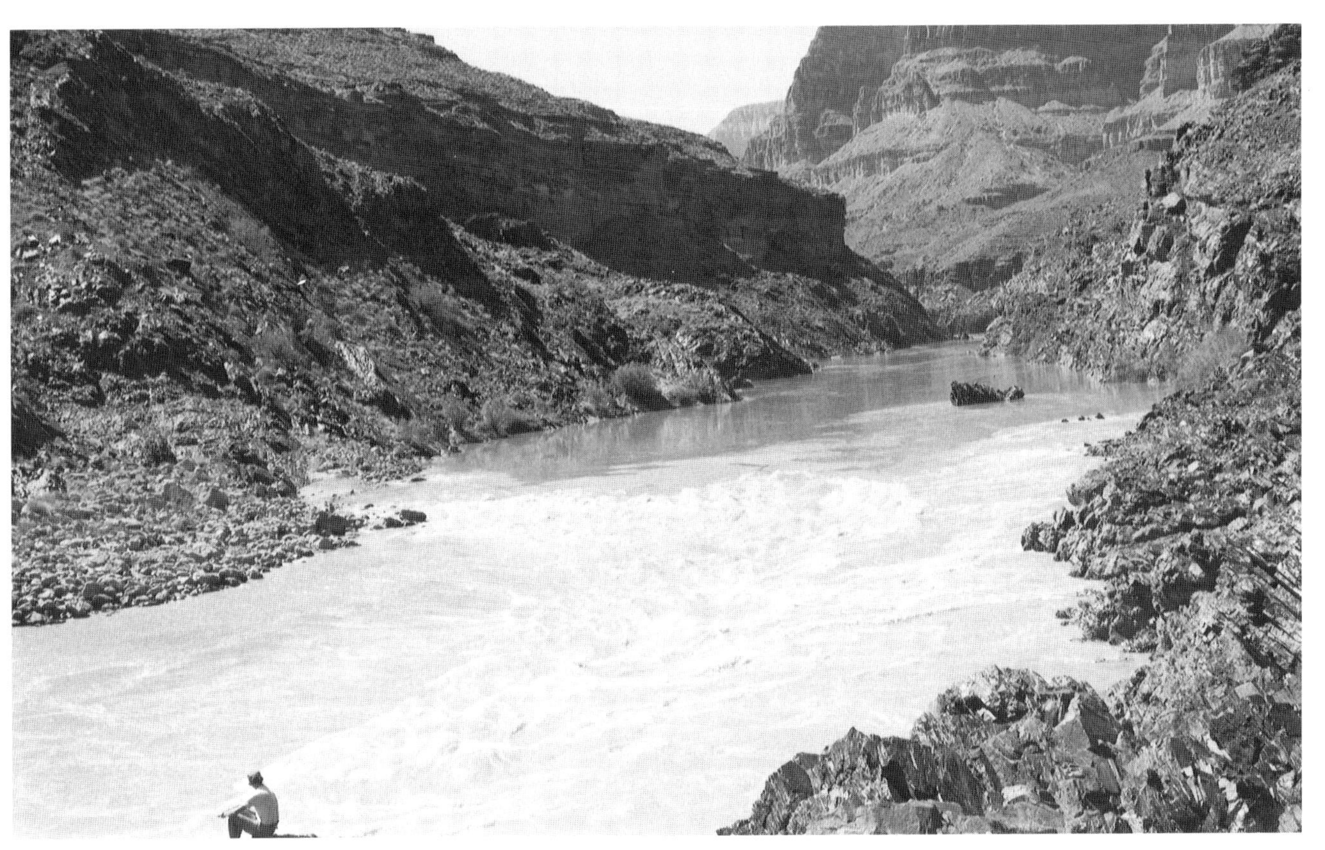

36. 128-Mile Rapid, mile 128.5,
September 17, 1991, 11:00 A.M.

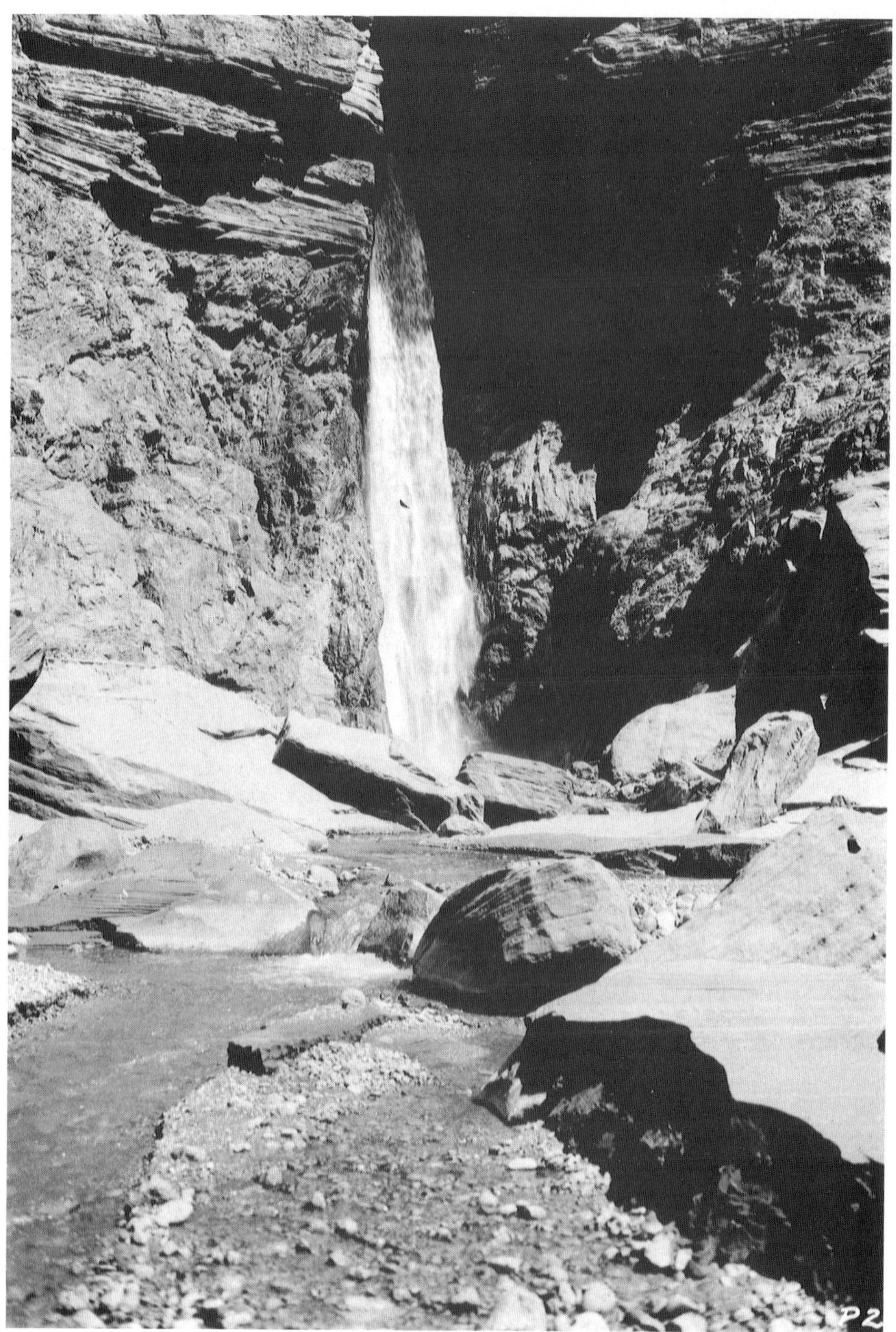
P2

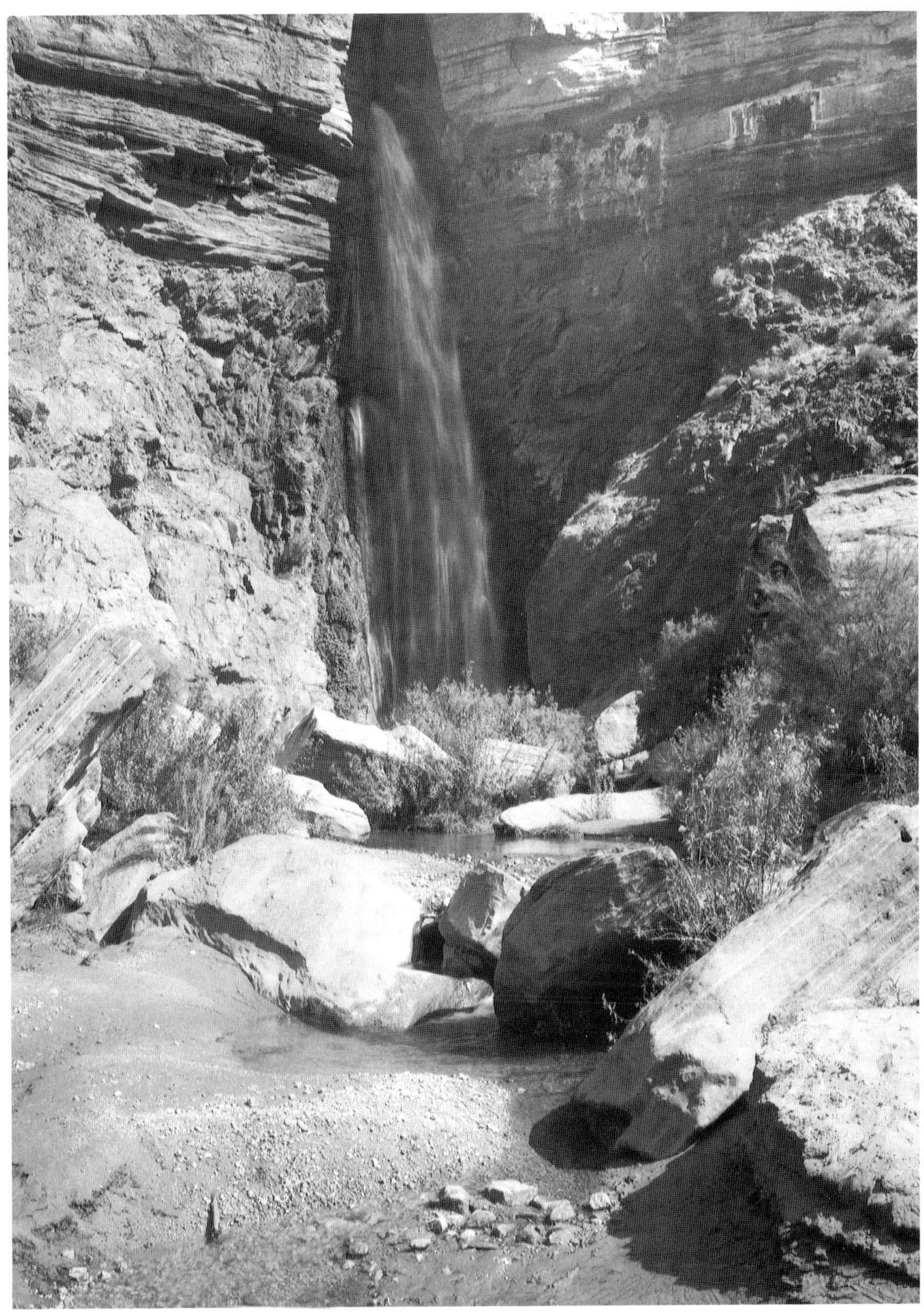

37. Deer Creek Falls, mile 136.3, September 19, 1991, 10:00 A.M.

37. Deer Creek Falls, mile 136.3, September 19, 1991, 10:00 A.M.

John Wesley Powell, in his 1875 account, wrote that Deer Creek Falls "leaps into the Colorado by a direct fall of more than a hundred feet, forming a beautiful cascade. . . . The stream pours through a narrow crevice above into a deep pool below. Around on the rocks, in the cave-like chamber, are set beautiful ferns, with delicate fronds and enameled stalks. . . . This delicate foliage covers the rocks all about the fountain, and gives the chamber great beauty. But we have little time to spend in admiration, so on we go." In 1923, R. C. Moore was more succinct: "Deer Ck falls about 100 ft over the granite."

Today, Deer Creek Falls, tumbling in from the north edge of the Colorado River, is a popular stop for visitors, many of whom hike the trail up Deer Creek Canyon.

Like so many of the other 1923 photographs, this one, probably by Freeman, taken on September 10, was made from atop a slippery, pointed boulder. The two views reveal that the rate of flow from the falls is much less today than it was in Birdseye's time. The edge of the sand bank in the foreground is much more sharply delineated and pristine in the older photo; its present state is the result of human foot traffic.

38. Kanab Creek, 1.5 miles up Kanab Creek from mile 143.5, September 19, 1991, 2:00 P.M.

Kanab Creek, the largest tributary flowing in from the north of Grand Canyon, extends upstream more than fifty miles, past the small town of Kanab, Utah. Powell's second expedition down the Colorado, in 1872, ended at the mouth of Kanab Canyon, where the crew met a resupply pack train and then followed the creek out of the gorge. Powell contended that the river was high and rising, making further travel dangerous. Later rephotography at known flow rates refutes that explanation.

The 1923 group reached this point on September 11; they surveyed several miles up Kanab Canyon, taking baths and shaving in Kanab Creek. Pollution of the tributary now discourages such luxuries.

Freeman took this photo in late morning from a ledge of Bright Angel Shale on the west side of Kanab Creek, looking downstream to the south. Moore drew a cross section here and described the geology, writing at length about the "snuff-brown" dolomite layer in the Bright Angel Shale. Perhaps this was the same unit that geologist G. K. Gilbert, during a trip up the canyon in the 1870s, referred to as "Old Snuffy."

Comparing the original and repeat photographs, one can see that the edges of the stream appear sandier in the older view, whereas pebbles and cobbles mantle the slick-rock streambed in the new photo. Many of the larger boulders on the opposite side of the creek have been rearranged by flash floods down this narrow side canyon.

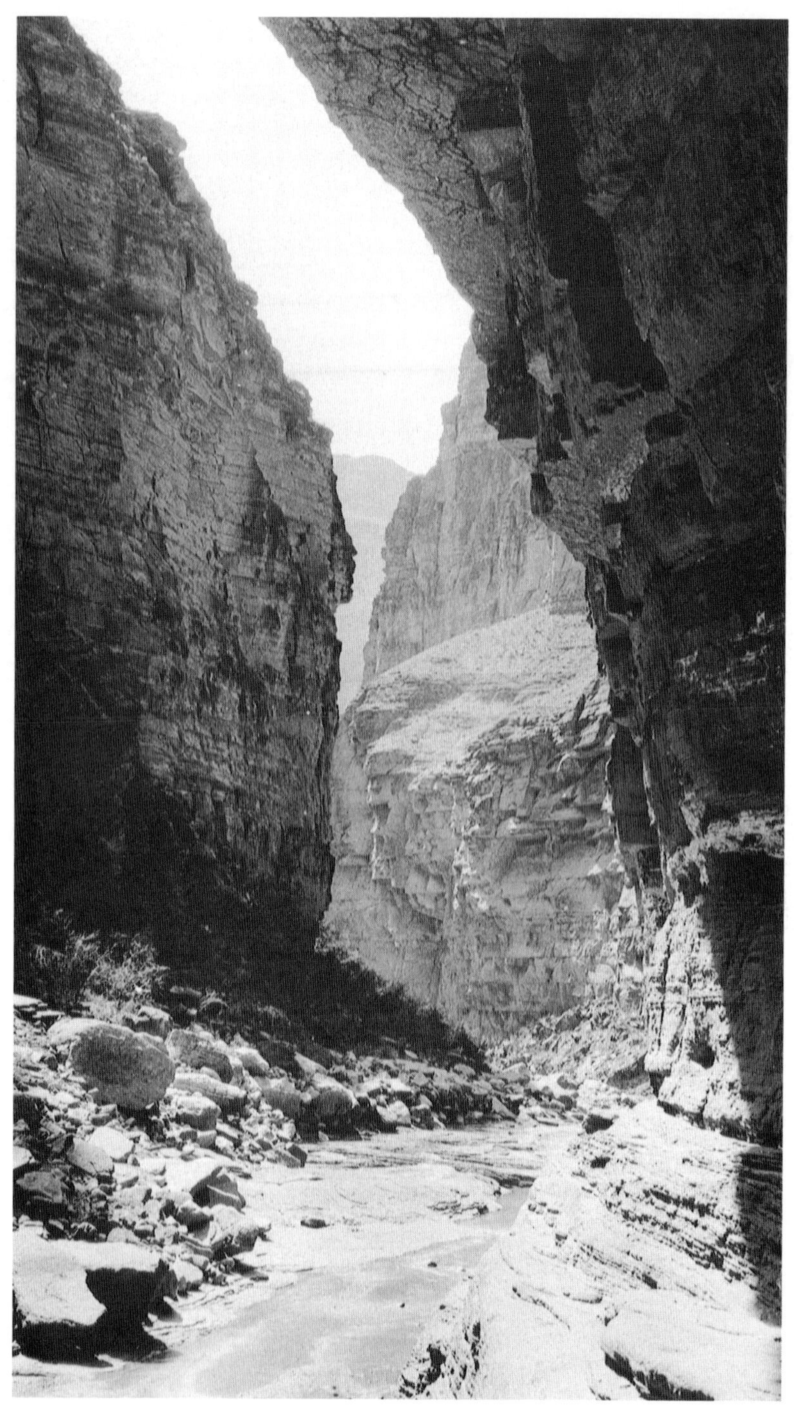

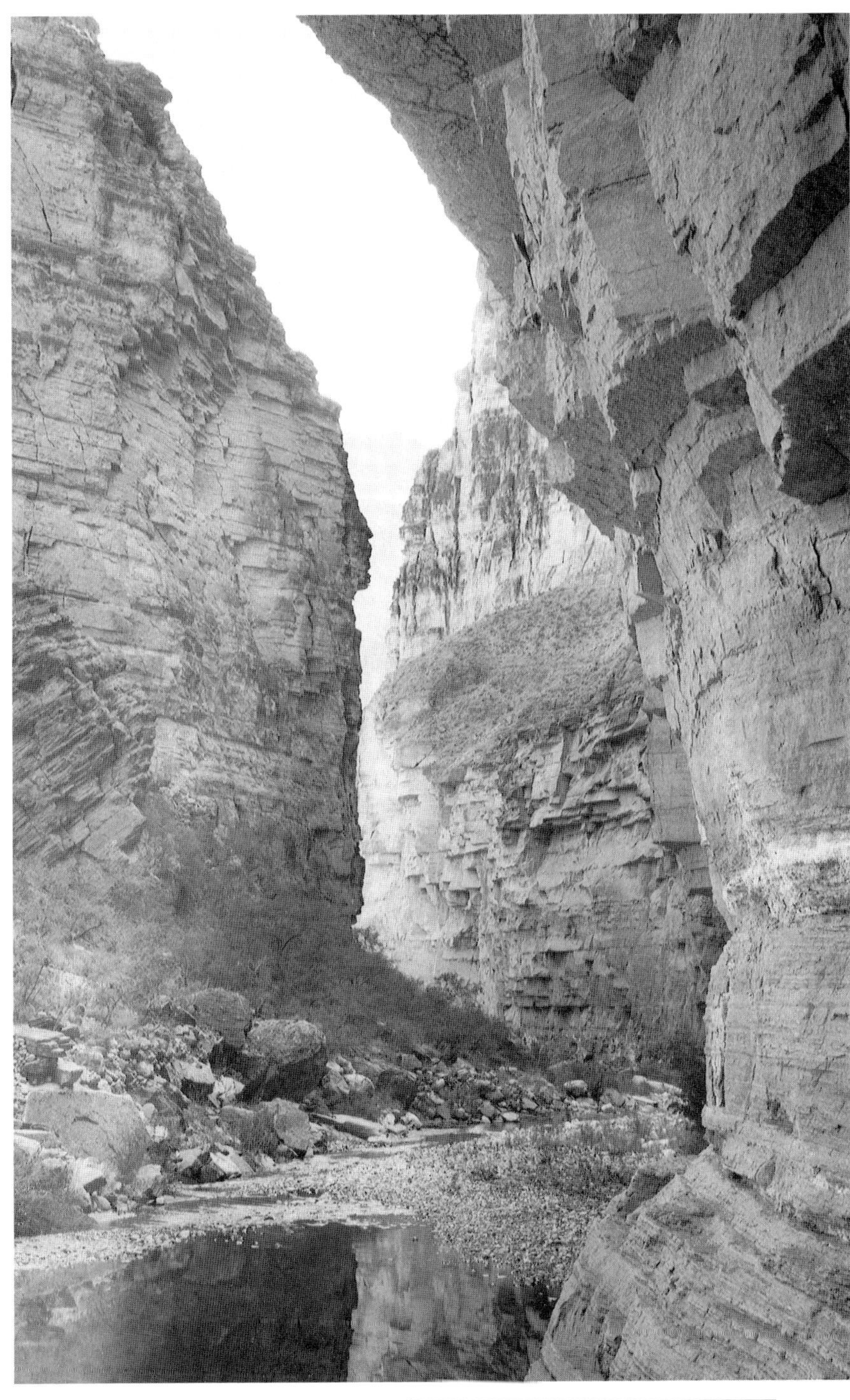

38. Kanab Creek, 1.5 miles up Kanab Creek from mile 143.5, September 19, 1991, 2:00 P.M.

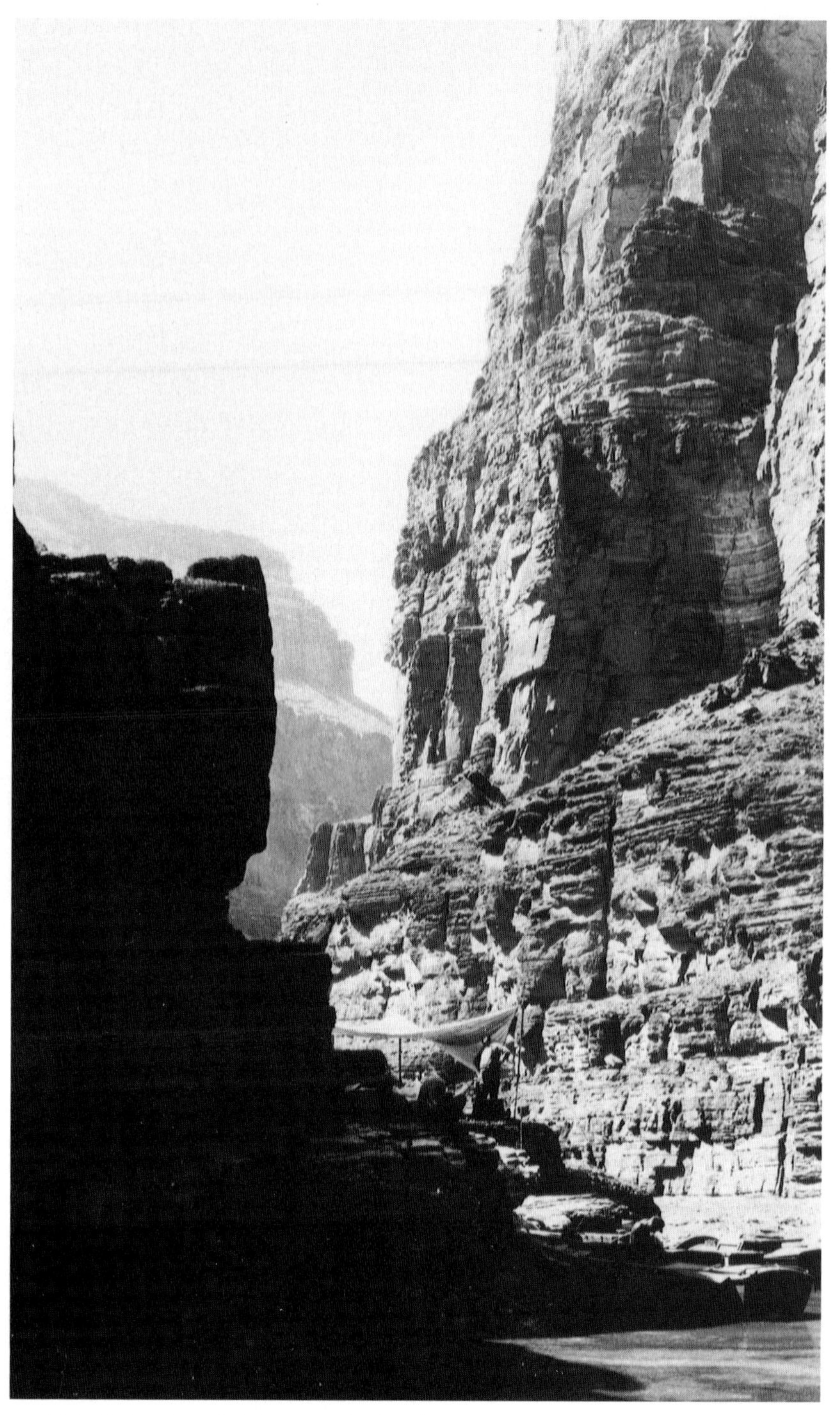

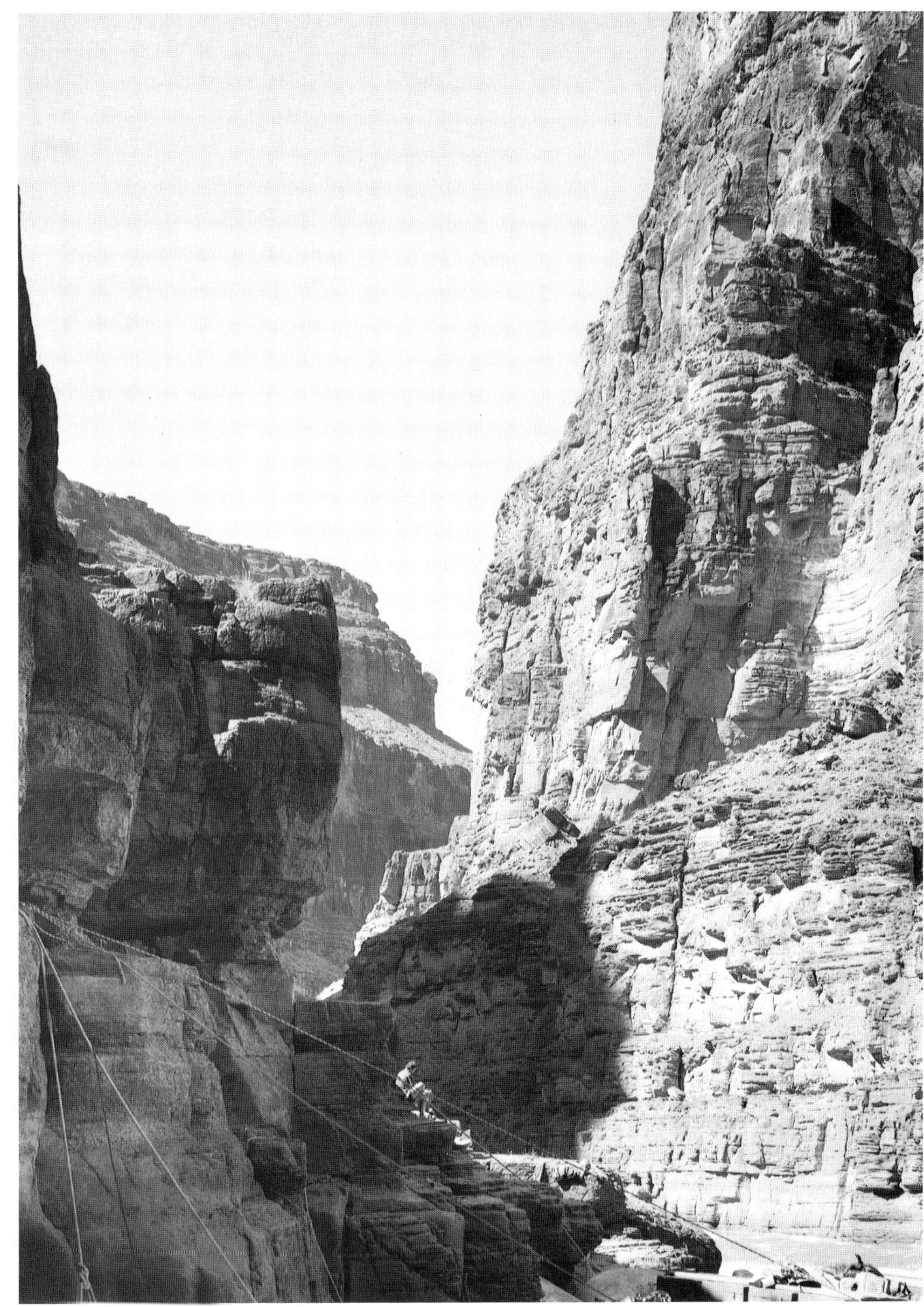

39. Mouth of Havasu Creek, mile 156.7, September 20, 1991, 10:30 A.M.

39. Mouth of Havasu Creek, mile 156.7, September 20, 1991, 10:30 A.M.

At the mouth of Havasu Creek, a popular stop for river travelers today, the 1923 crew was met by nineteen Havasupai Indians bearing supplies brought down from the South Rim via the village of Havasupai. Birdseye also brought in a new cook, Felix Kominsky, whom Blake described as "somewhat of a wonder. He made a cake in a dutch oven and never got out of humor when the wind blew sand and ashes into everything." The 1923 crew set up shop here on a broad, flat rock bench at the mouth of the spring-fed tributary.

This view—the original was probably taken in early afternoon on September 13, 1923, by LaRue—faces south, or downstream, from the mouth of Havasu Creek. The water there was shallow, about three feet deep. Blake's diary recalls: "We turned the boat crossways of the stream and let the current wedge it between the walls where it lay motionless while we wrote letters."

Our repeat photograph was the only one we shot from the boat, and we used a similar tactic—wedging the bow and stern of our boat crosswise between the cliffs at the mouth of the creek. We then set up the camera over the motor well in the stern. The tie-up ropes of several other boats stopped here are visible in the new photograph.

40. Above National Canyon, mile 166.5, September 20, 1991, 3:00 P.M.

This scene was originally photographed in early afternoon, probably on September 15, 1923, from the south side of the river looking downstream, or west, toward the mouth of National Canyon just inside the Hualapai Indian Reservation. The vantage point, a bank about thirty feet above the river covered with sharp, blocky limestone and shale boulders, was memorable because in 1991 it was the home of a pink Grand Canyon rattlesnake.

In the new photograph, vegetation—mostly tamarisks—obscures the foreground and decorates the gravel bar below. The large sandbar visible near the tied-up boat in the older photo is now covered with vegetation and eroded down to its cobble floor.

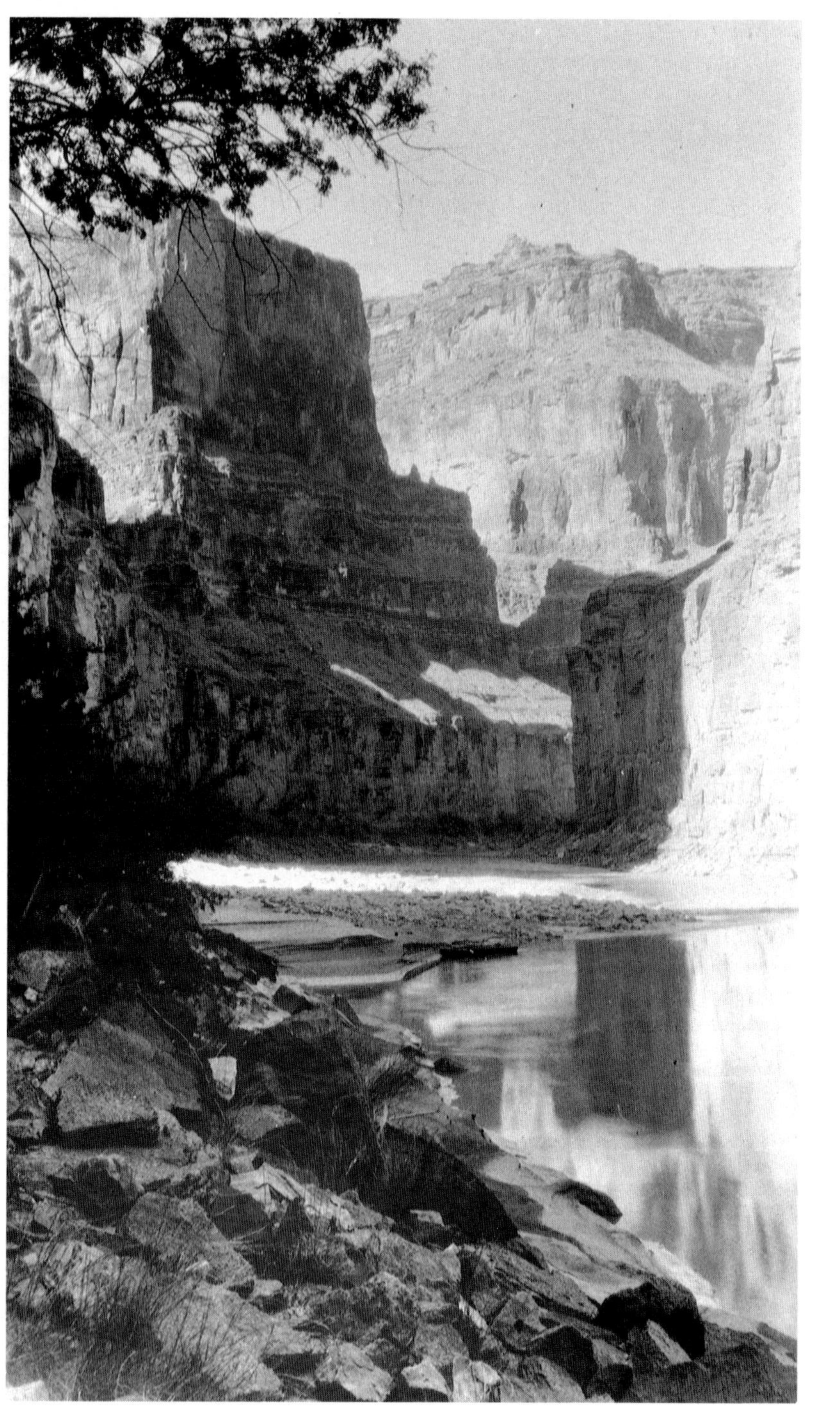

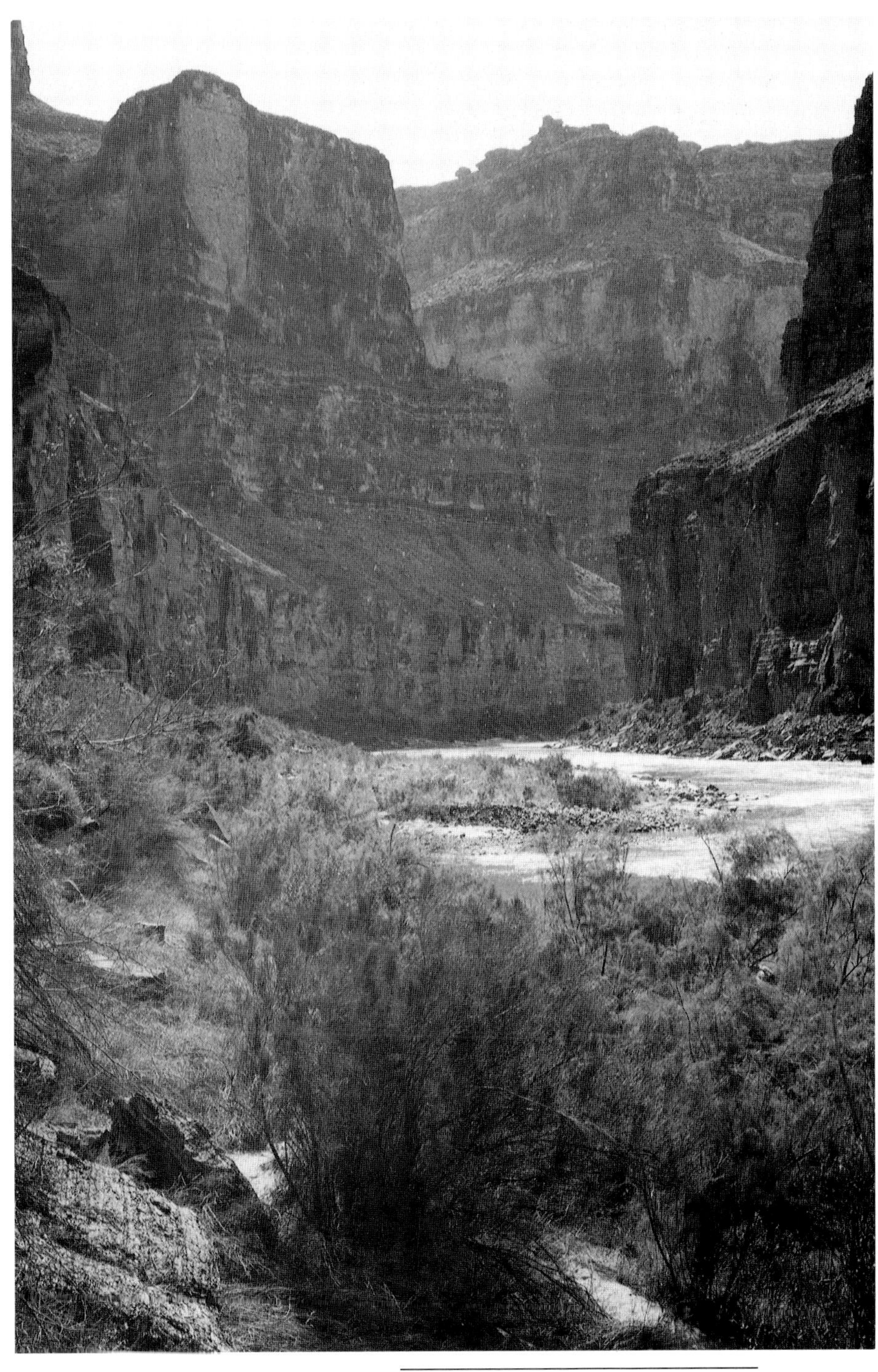

40. Above National Canyon, mile 166.5, September 20, 1991, 3:00 P.M.

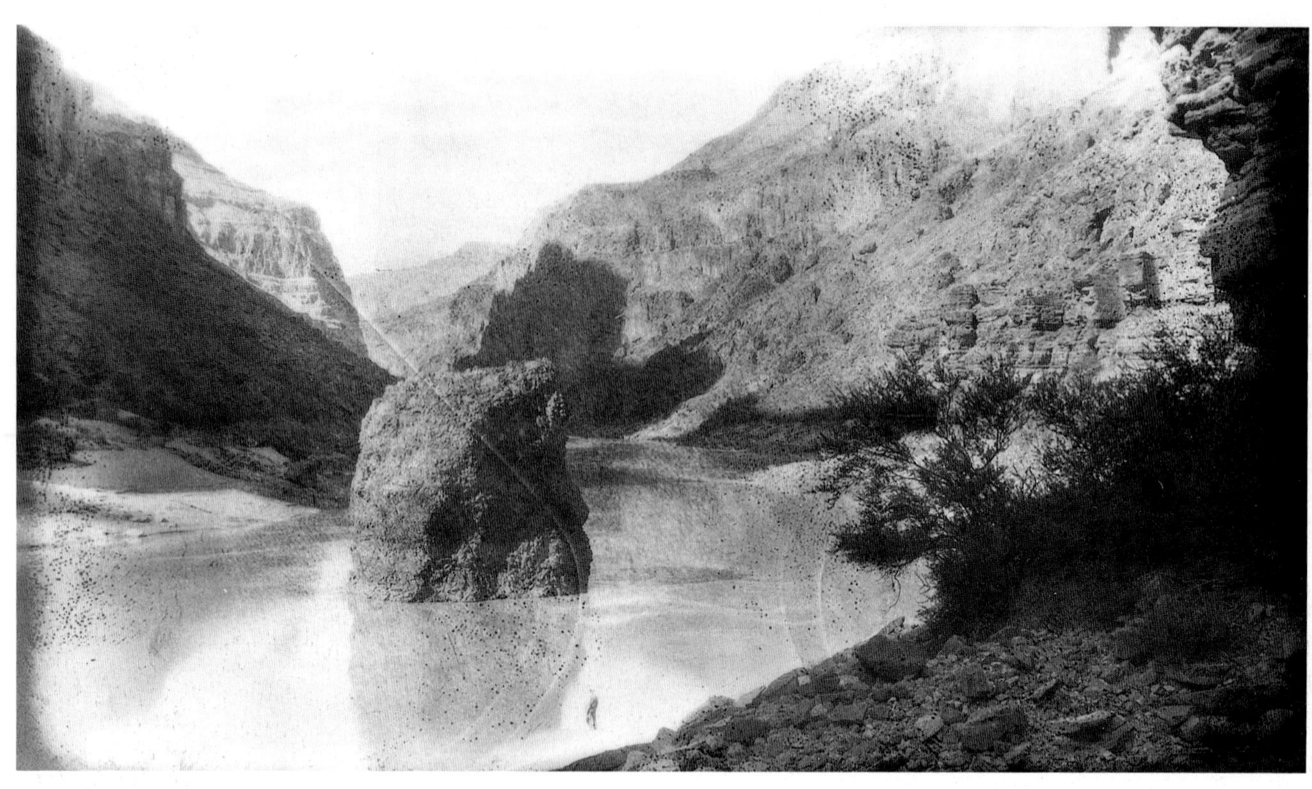

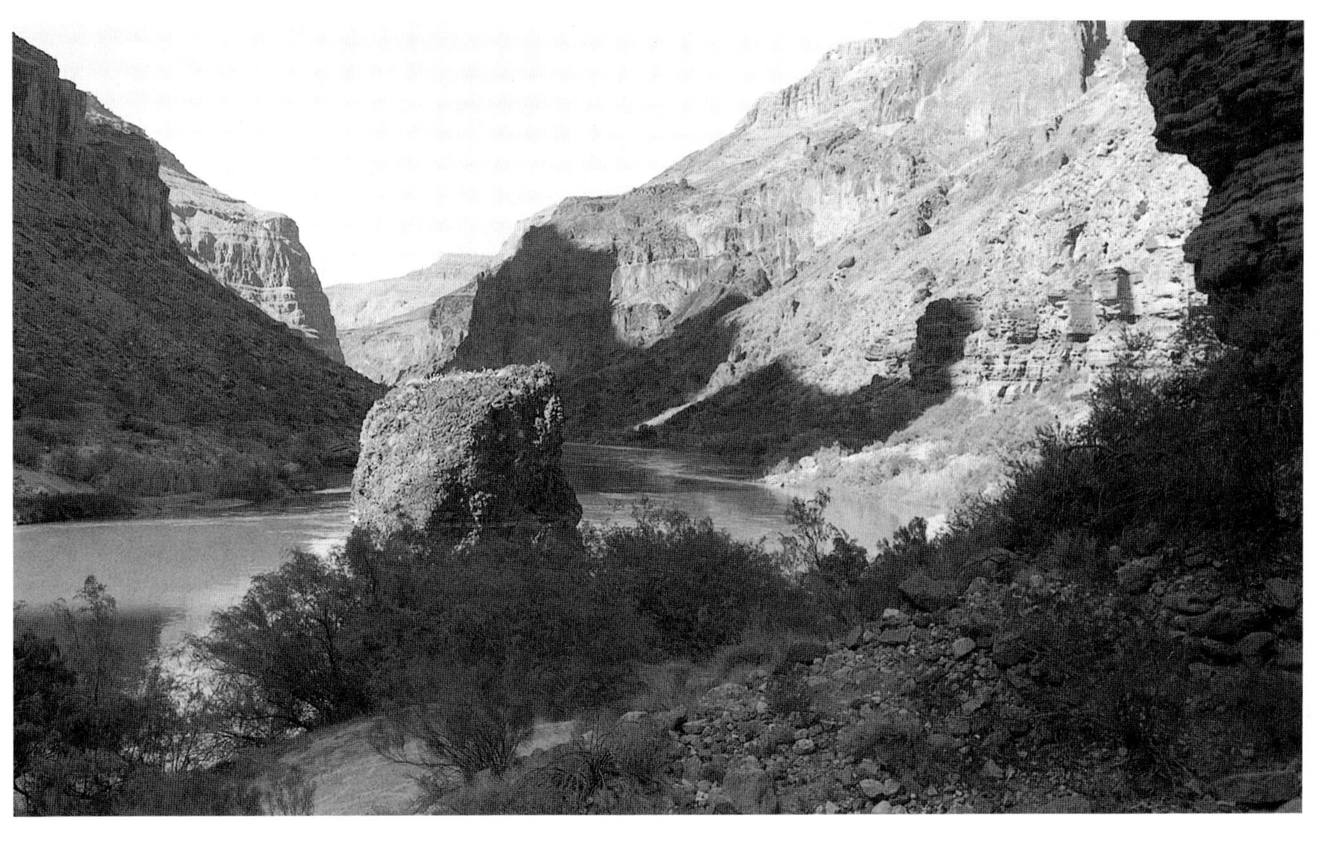

41. Vulcan's Anvil, mile 177.9,
September 21, 1991, 10:00 A.M.

41. Vulcan's Anvil, mile 177.9, September 21, 1991, 10:00 A.M.

Vulcan's Anvil (also known as Vulcan's Forge and labeled "Lava Pinnacle" on the 1923 survey map) is the black basalt neck of a volcano that erupted about 1.2 million years ago. The eruption, along with numerous others in the vicinity, dammed the river with lava until eventually the lake that formed behind the dams cut through and washed away everything but the neck that remains.

R. C. Moore had a slightly different interpretation of Vulcan's Anvil. He described it in his notebook as "a striking little islet of basalt ±100 ft in width, nearly circular and with vertical sides. It rises above higher water mark as shown by the growth of cactus and bushes on its top. This lava islet is apparently in place and I interpret as a remnant of the upstream extension of the main lava cascade near Lava Falls."

Emery Kolb took the original photo in mid-morning on September 18, 1923, looking downstream, or generally southwest, from the north side of the river, three thousand feet below Toroweap Point. The vantage point was a layer of sandstone in the Bright Angel Shale; we found it largely removed by erosion and the photo difficult to replicate.

The quality of the older print is poor, with several lines and overexposed areas like that at the upper right. A person standing on a small sandbar at river's edge, however, is visible for scale in the lower center of the picture. Our repeat photograph shows the ubiquitous tamarisk and other recently introduced vegetation covering much of the foreground and what was once a sandy beach on the far left of the older photo.

42. Lava Falls, mile 179.5, September 21, 1991, 11:00 A.M.

Few spots on the river are as feared by river runners as Lava Falls. It is among the most dramatic drops in the river, some thirty-seven feet, and one of the most challenging rapids on the Colorado.

The 1923 crew had its own set of problems at Lava Falls. On the night of September 18, floodwaters moving in from the Little Colorado pushed the river level up twenty feet. The crew got up several times during the night to pull their boats to higher ground to protect them from pounding by the swollen river. They spent September 19 watching water flood through the rapids ("No rocks are showing at all," wrote Blake) and waiting for the water to go down. By the time the water had receded enough that they could renew the trip, the boats were left high above the river and had to be skidded down. The delay made the crew late to meet a resupply outfit at Diamond Creek, sparking rumors that the expedition had been lost.

With time provided by the flood, Moore made copious notes, including remarks about the geology visible in these photos. "In the creek which enters Colorado River on the south opposite Lava Falls there is a very large accumulation of alluvial debris, large boulders, etc." The 1923 photo was taken from the edge of the alluvial fan to which Moore refers, at the mouth of Prospect Canyon. He continued: "450–500′ above the river opposite Lava Falls, on north side river is a lenticular mass of rounded boulders, mostly of lava but including one large ls [limestone] boulder 10–15′ dia, which marks base of one of the flows of lava." This and another lava flow, downstream in the middle distance, are visible in these photographs.

The 1923 photo in this pair was probably taken by LaRue in mid-morning, probably on September 20, looking downstream to the west above Lava Falls. Although the angles on distant features appear to be identical, the foreground in the two photos is quite different. It looks as if flash flooding has completely altered the mouth of Prospect Canyon since 1923, apparently depositing a new boulder fan from which erosion later carved a new wash, now lined with tamarisks.

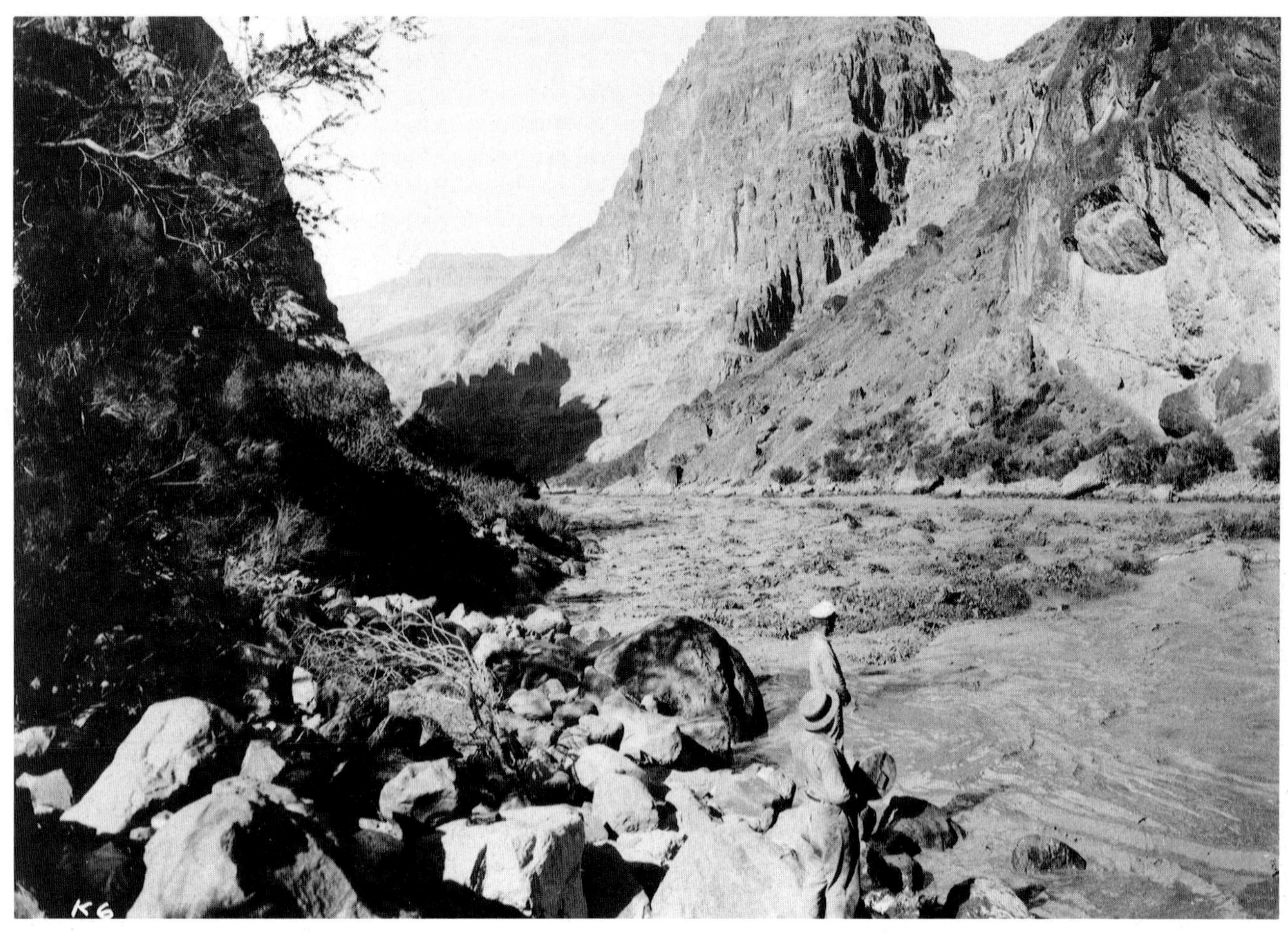
K6

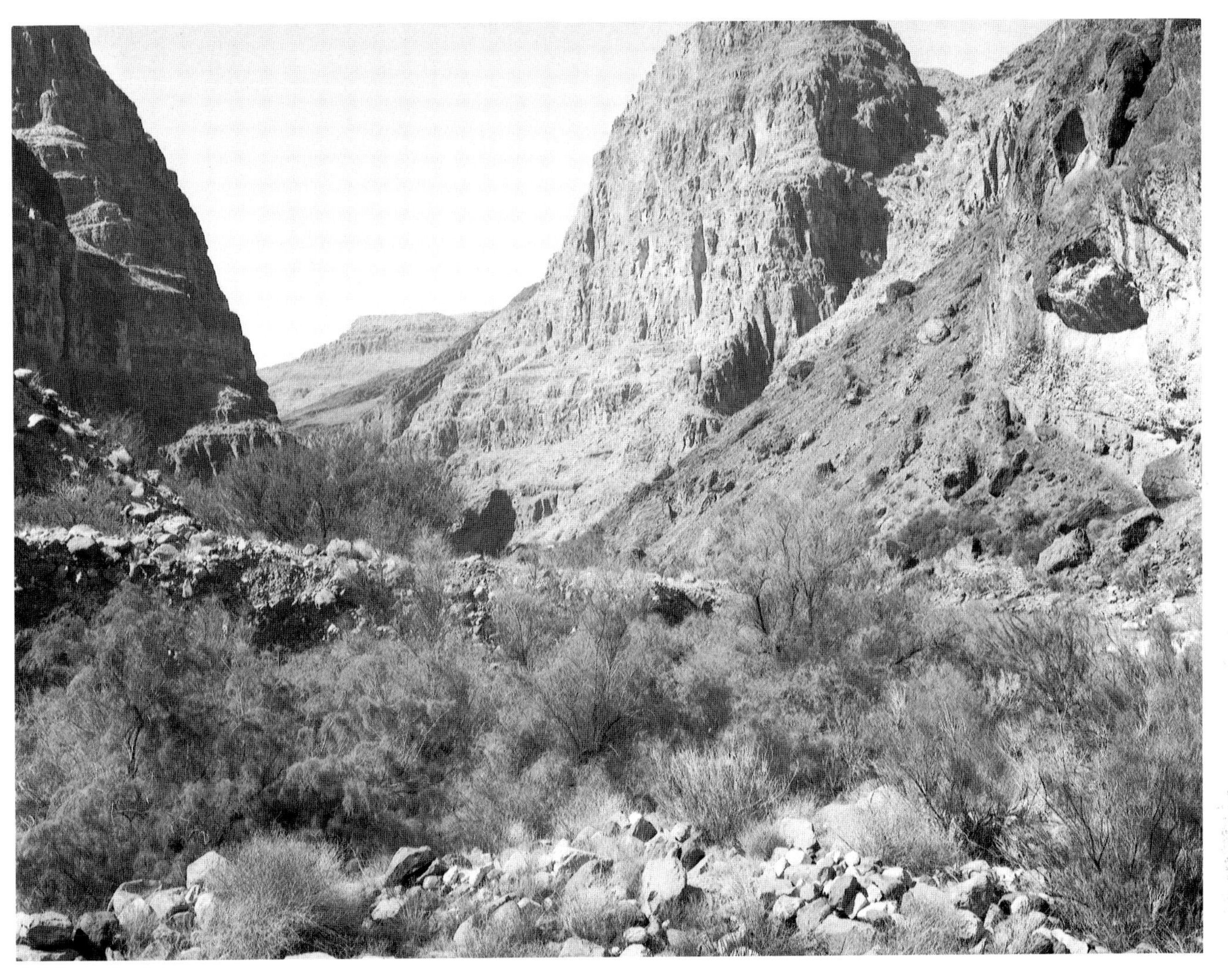

42. Lava Falls, mile 179.5,
September 21, 1991, 11:00 A.M.

43. Above Diamond Peak,
mile 222.5, September 22, 1991,
11:00 A.M.

Near the confluence of Diamond Creek and the Colorado River stands Diamond Peak, a residual mountain capped by Redwall Limestone. This view of Diamond Peak was first taken in mid-morning sometime between October 2 and 6, 1923, from about one-quarter mile above the east bank of the river, looking southwest. The vantage point was a large, polished boulder of lava.

Comparing the older photo with our repeat version shows the sparsely vegetated dune and shoreline in the middle distance now partially overgrown and stabilized by plant life. We located the ocotillo in the lower center of the old photo; it has since died and its skeletal remains are obscured in the new photo by the brushy growth at lower right. The nature of the plant community above the high-water line, however, has not changed since 1923.

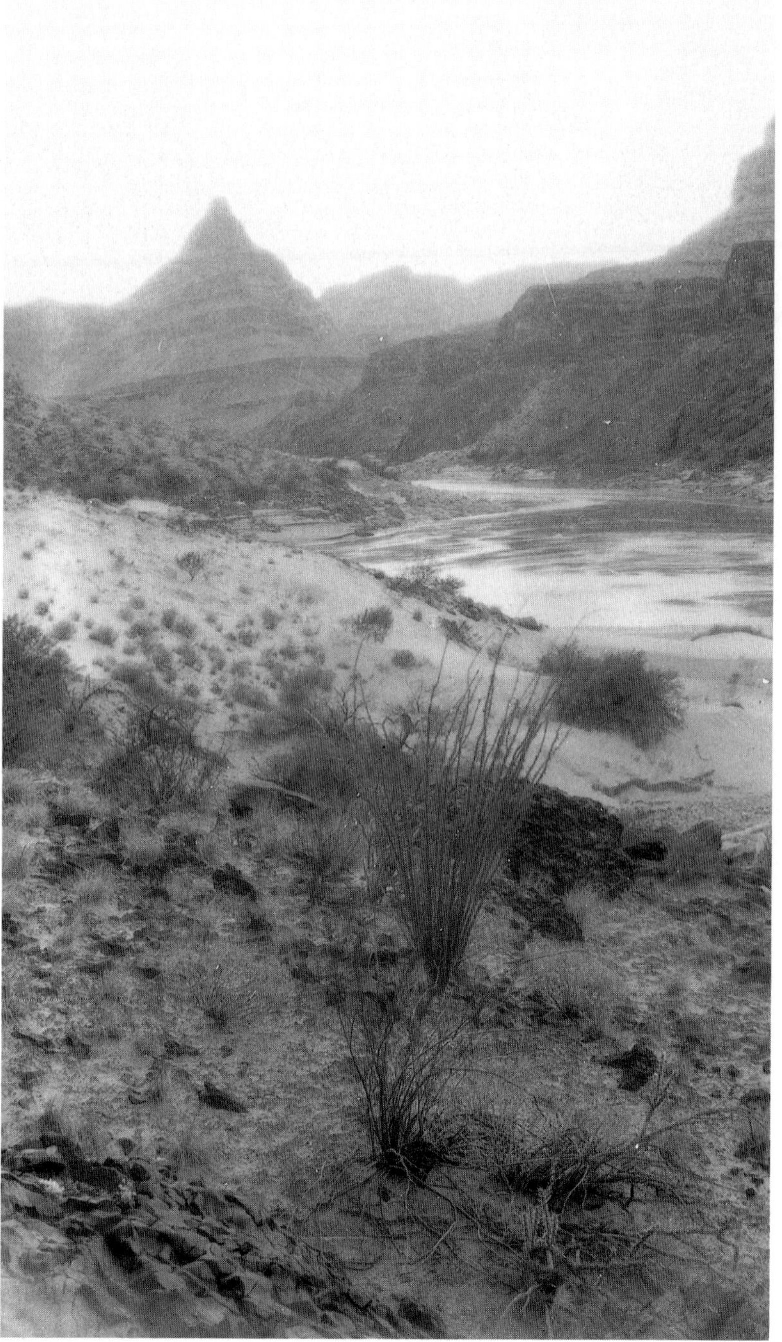

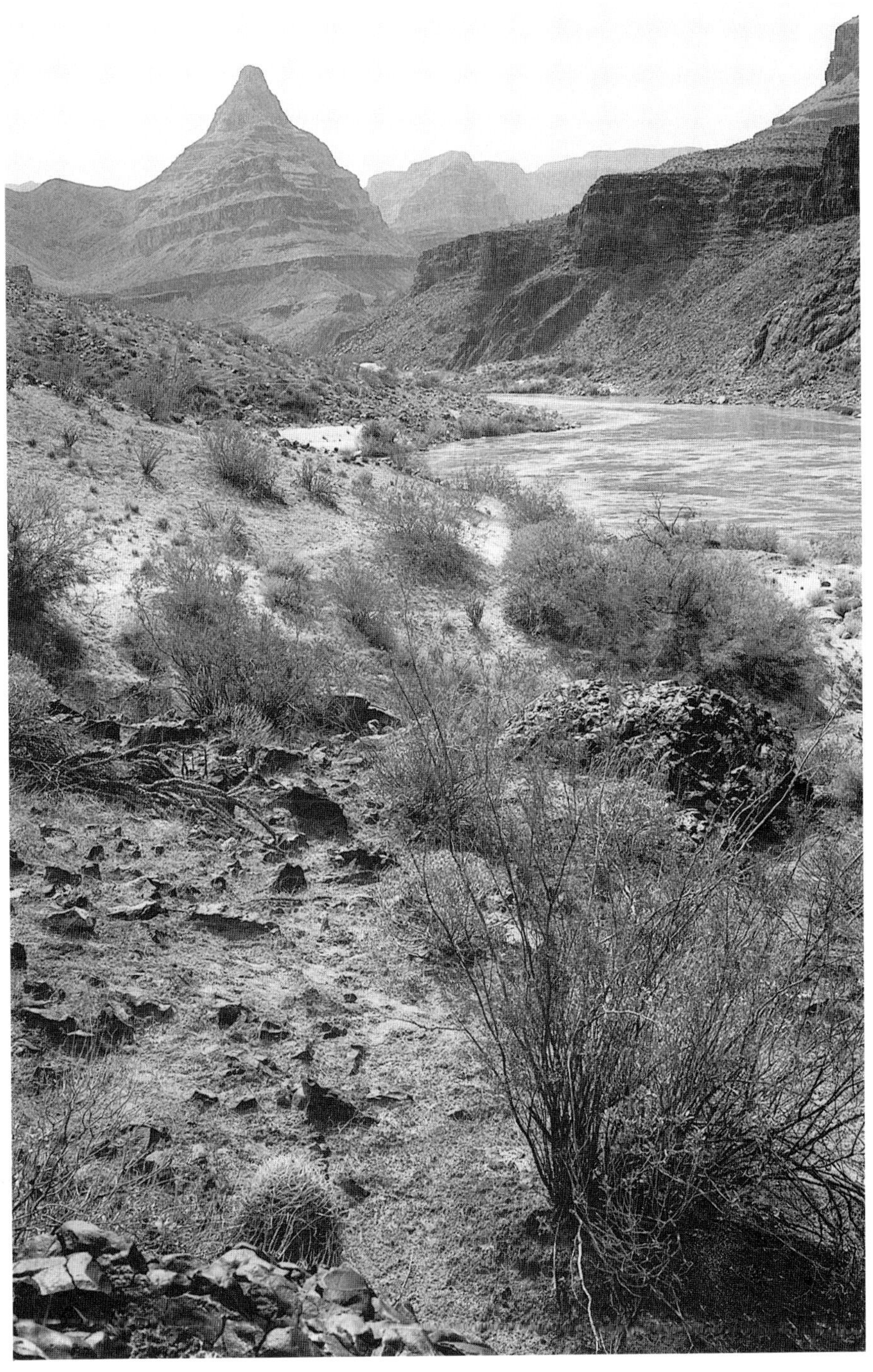

43. Above Diamond Peak, mile 222.5, September 22, 1991, 11:00 A.M.

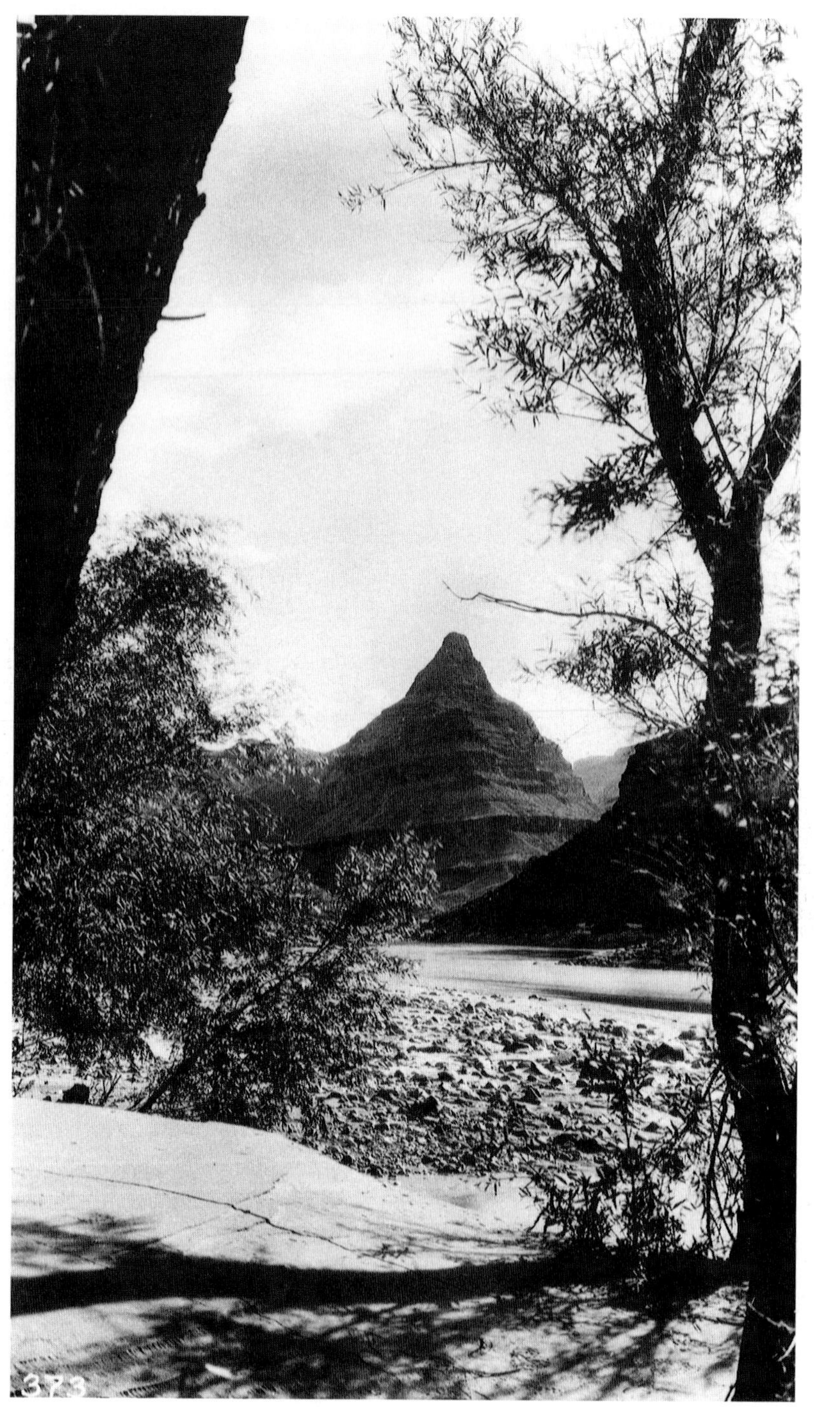
373

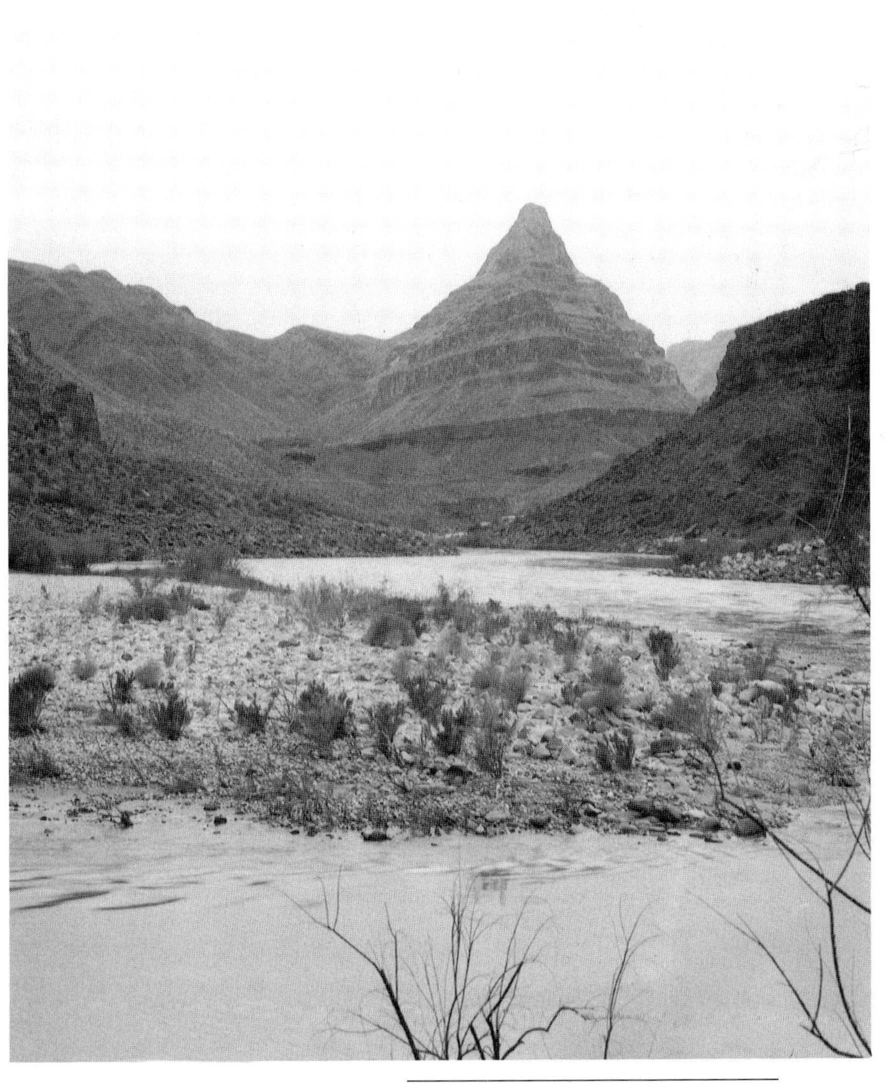

44. Diamond Peak, mile 222.8, September 22, 1991, 1:00 P.M.

44. Diamond Peak, mile 222.8, September 22, 1991, 1:00 P.M.

This 1923 view of Diamond Peak, probably taken by Freeman sometime between October 2 and 6, faces mostly south from a sandbar on the east side of the river. Most of the old vegetation, primarily willow, is gone today, and new stands of tamarisk made rephotography difficult. The sandbar in the photo's far foreground is much the same, although it sustains new vegetation. The sand in the near foreground has been partially rearranged by river flow, and today this sand bank is completely overgrown with tamarisks and other plants.

45. 224-Mile Rapid, mile 223.3, September 22, 1991, 3:00 P.M.

Birdseye's group surveyed a dam site below Diamond Peak, and Moore made a rough sketch of the rock formations and faulting here. They also smashed the stern of the *Grand* into a rock on the west side of 224-Mile Rapid.

Their photo was taken sometime between October 2 and 6, 1923, from a boulder about a hundred feet above the west bank, looking mostly south. The defile known as 224-Mile Canyon enters the Colorado at the base of Diamond Peak. The beach at the mouth of this canyon is covered with considerably more vegetation in the newer photo, and much of its sand is now washed away. Vegetation in the foreground is reduced in the newer photo, although the variety of plants remains the same. Many of the ocotillos in the older photo's foreground still thrive today. In general, vegetation above the high water line has changed little since 1923.

The rocks in the lower left corner line the edge of a small gully, and many appear to have been rearranged, probably by local rainfall and runoff. A small boulder with two lines crossing to form a + in the middle foreground is still in place and was used in relocating the vantage point.

This, the last day of our expedition, was continuously threatened by dark skies and the rumbling of thunder in the canyon. During the evening—as Powell once wrote—"It rains."

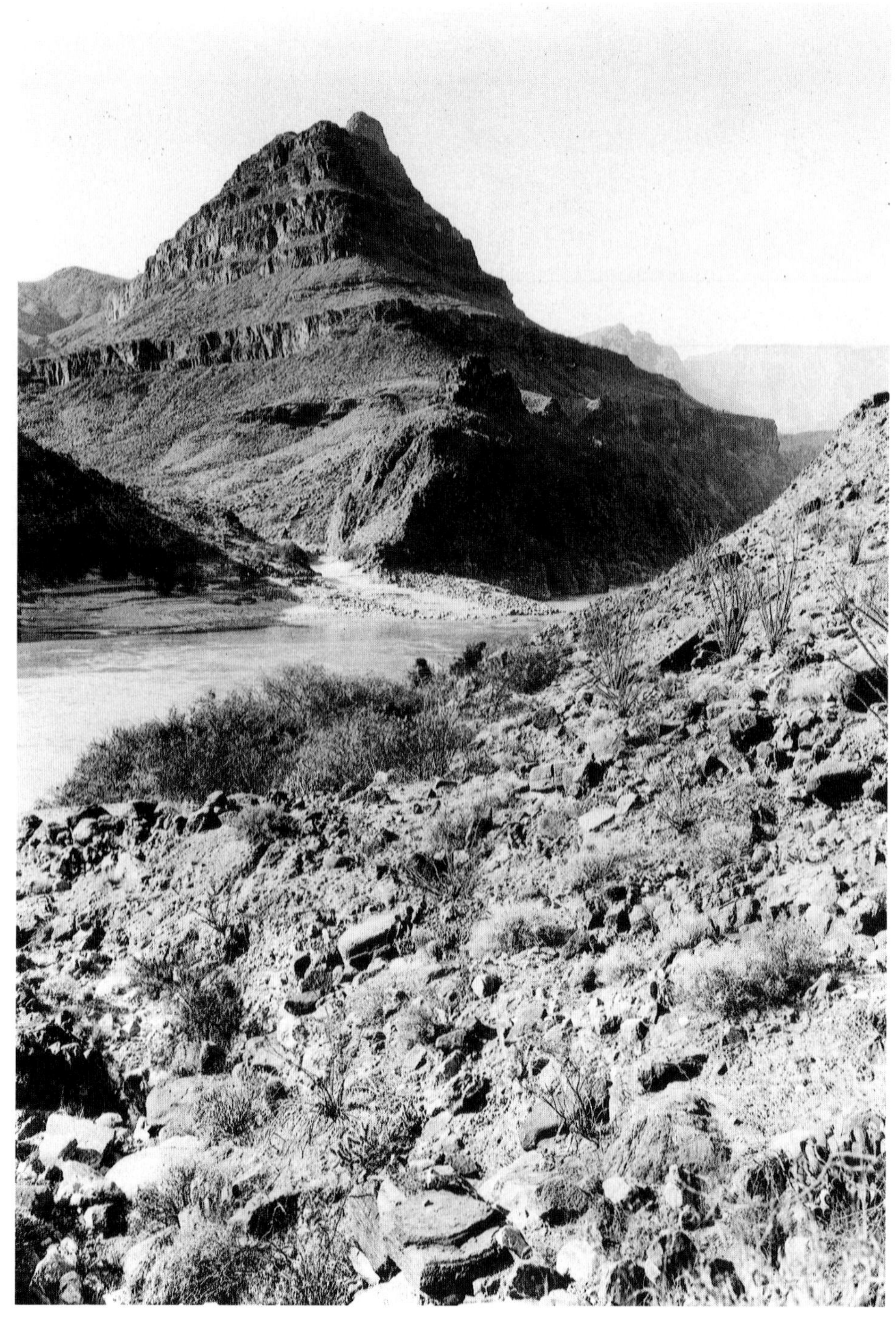

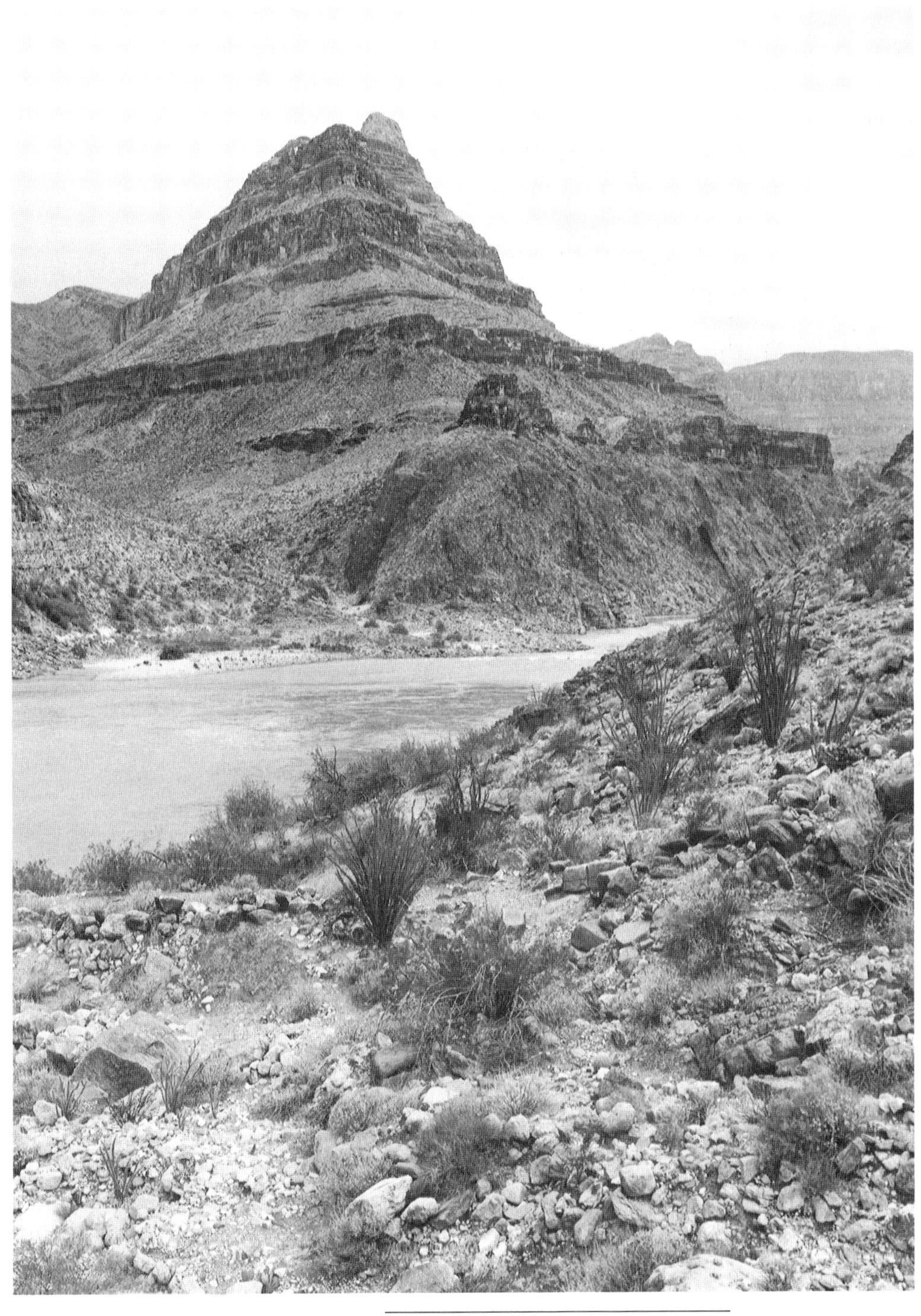

45. 224-Mile Rapid, mile 223.3, September 22, 1991, 3:00 P.M.

PHOTOGRAPH LOCATIONS

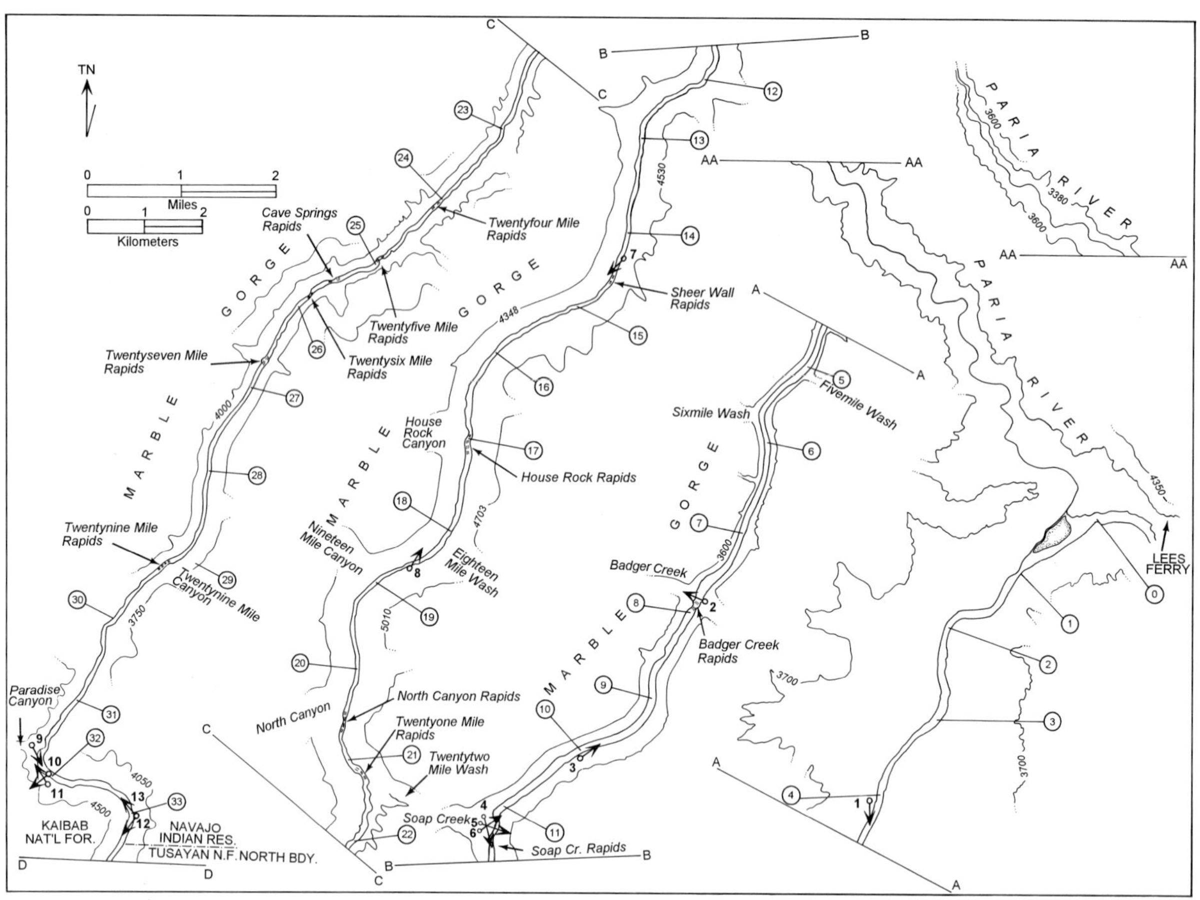

1. Navajo Bridge, mile 4.1, September 24, 1991, 9:00 A.M.

2. Badger Creek Rapid, mile 7.8, September 8, 1991, noon.

3. 10-Mile Rock, mile 10.0, September 8, 1991, 3:00 P.M.

4. Canyon wall near Soap Creek, mile 11.1, September 9, 1991, 8:30 A.M.

5. Soap Creek Rapid, mile 11.2, September 9, 1991, 10:00 A.M.

6. Soap Creek Rapid, mile 11.2, September 9, 1991, 1:00 P.M.

7. Near Tanner Wash (Sheerwall Rapid), mile 14.4, September 9, 1991, 3:30 P.M.

8. House Rock, mile 18.6, September 9, 1991, 6:00 P.M.

9. Above Vasey's Paradise, mile 31.6, September 10, 1991, 10:00 A.M.

10. Vasey's Paradise from Boulder Island, mile 31.9, September 10, 1991, noon.

11. Stanton's Cave, mile 32.0, September 10, 1991, 3:00 P.M.

12. Redwall Cavern, mile 33.1, September 10, 1991, 4:00 P.M.

13. Redwall Cavern, view upstream, mile 33.1, September 10, 1991, 5:00 P.M.

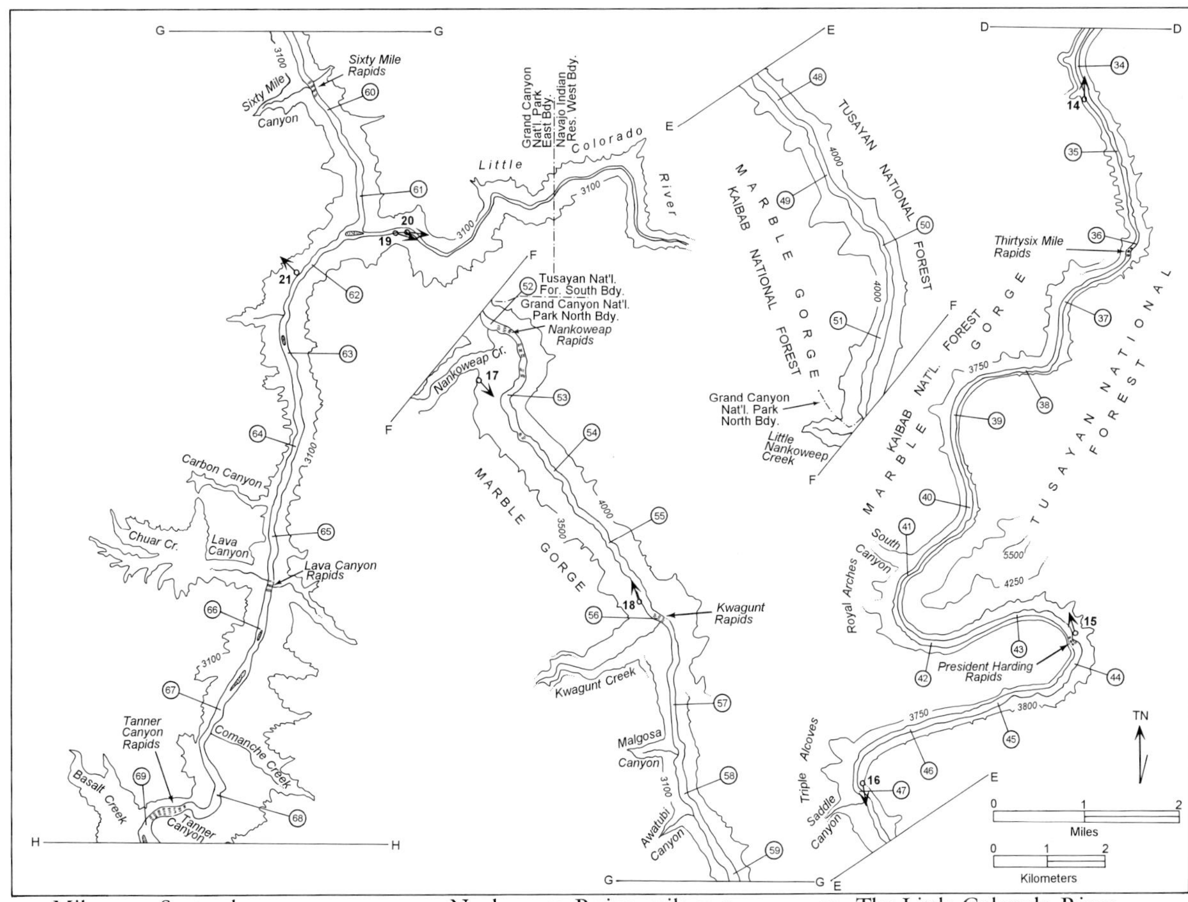

14. Mile 34.4, September 10, 1991, 6:00 P.M.

15. Camp at President Harding Rapid, mile 43.6, September 11, 1991, 10:00 A.M.

16. Near Saddle Canyon, mile 46.9, September 11, 1991, 11:30 A.M.

17. Nankoweap Ruins, mile 53.0, September 11, 1991, 3:00 P.M.

18. Near Kwagunt Rapid, mile 55.8, September 12, 1991, 10:00 A.M.

19. The Little Colorado River, 0.5 miles up the Little Colorado from mile 61.5, September 12, 1991, noon.

20. The Little Colorado River, about 0.5 miles up the Little Colorado from mile 61.5, September 12, 1991, 1:30 P.M.

21. Hollows in the Tapeats Sandstone, mile 62.2, September 12, 1991, 4:00 P.M.

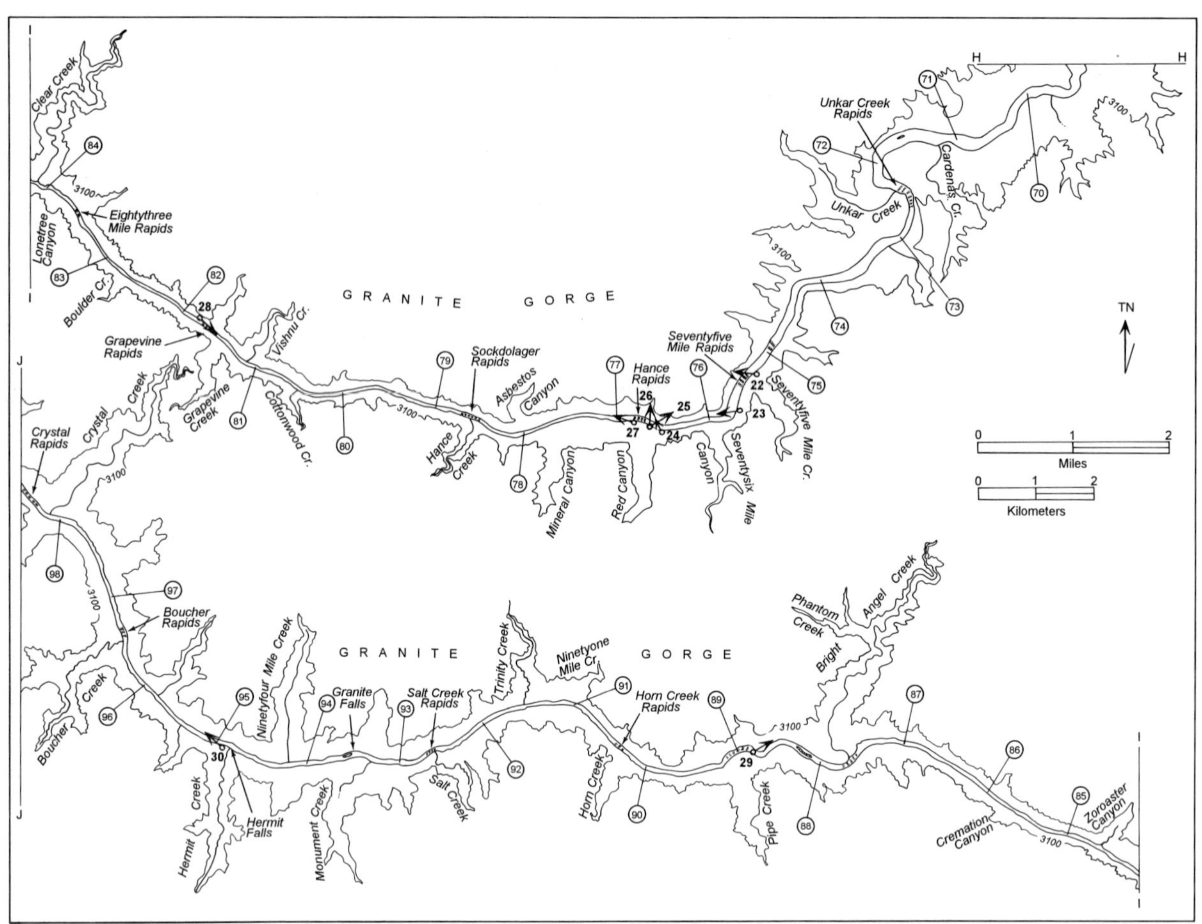

22. Nevills Rapid, mile 75.3, September 13, 1991, 1:00 P.M.

23. Algonkian Gorge above Hance Rapid, mile 75.7, September 13, 1991, 3:00 P.M.

24. Top Of Hance Rapid, mile 76.5, September 13, 1991, 5:00 P.M.

25. Looking upstream at Hance Rapid, mile 76.7, September 14, 1991, 11:00 A.M.

26. Looking across the river at Hance Rapid, mile 76.7, September 14, 1991, 11:00 A.M.

27. Foot of Hance Rapid, mile 76.8, September 14, 1991, 10:00 A.M.

28. Grapevine Rapid, mile 81.8, September 14, 1991, 2:00 P.M.

29. Camp at the foot of Bright Angel Trail, mile 88.9, September 14, 1991, 5:00 P.M.

30. Hermit Rapid, mile 94.9, September 15, 1991, 10:00 A.M.

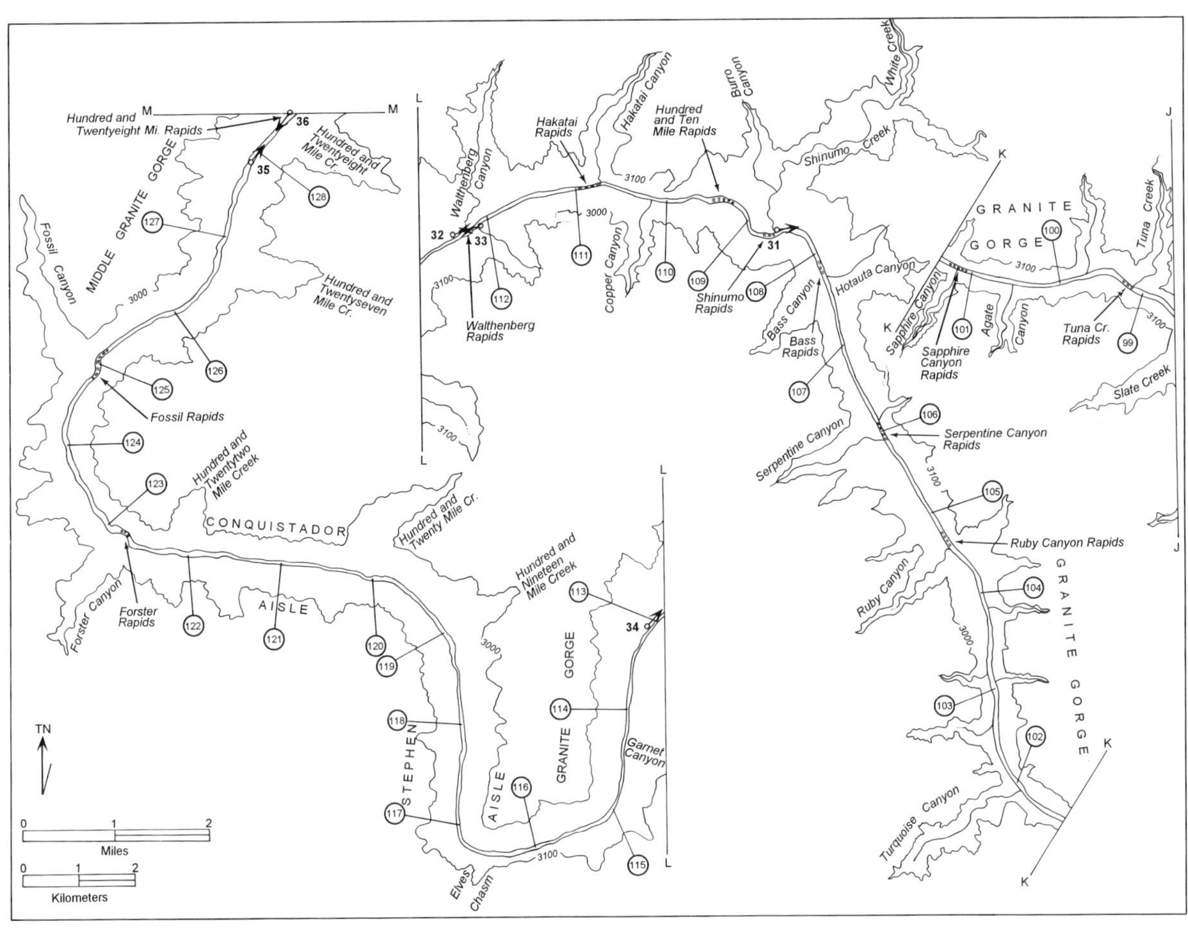

31. Shinumo Creek, mile 108.6, September 15, 1991, 3:00 P.M.

32. Above Waltenberg Rapid, mile 112.1, September 16, 1991, 10:00 A.M.

33. High above Waltenberg Rapid, mile 112.3, September 16, 1991, noon.

34. Below Waltenberg Rapid, mile 113.1, September 16, 1991, 3:00 P.M.

35. Below Fossil Rapid, mile 127.0, September 17, 1991, 9:00 A.M.

36. 128-Mile Rapid, mile 128.5, September 17, 1991, 11:00 A.M.

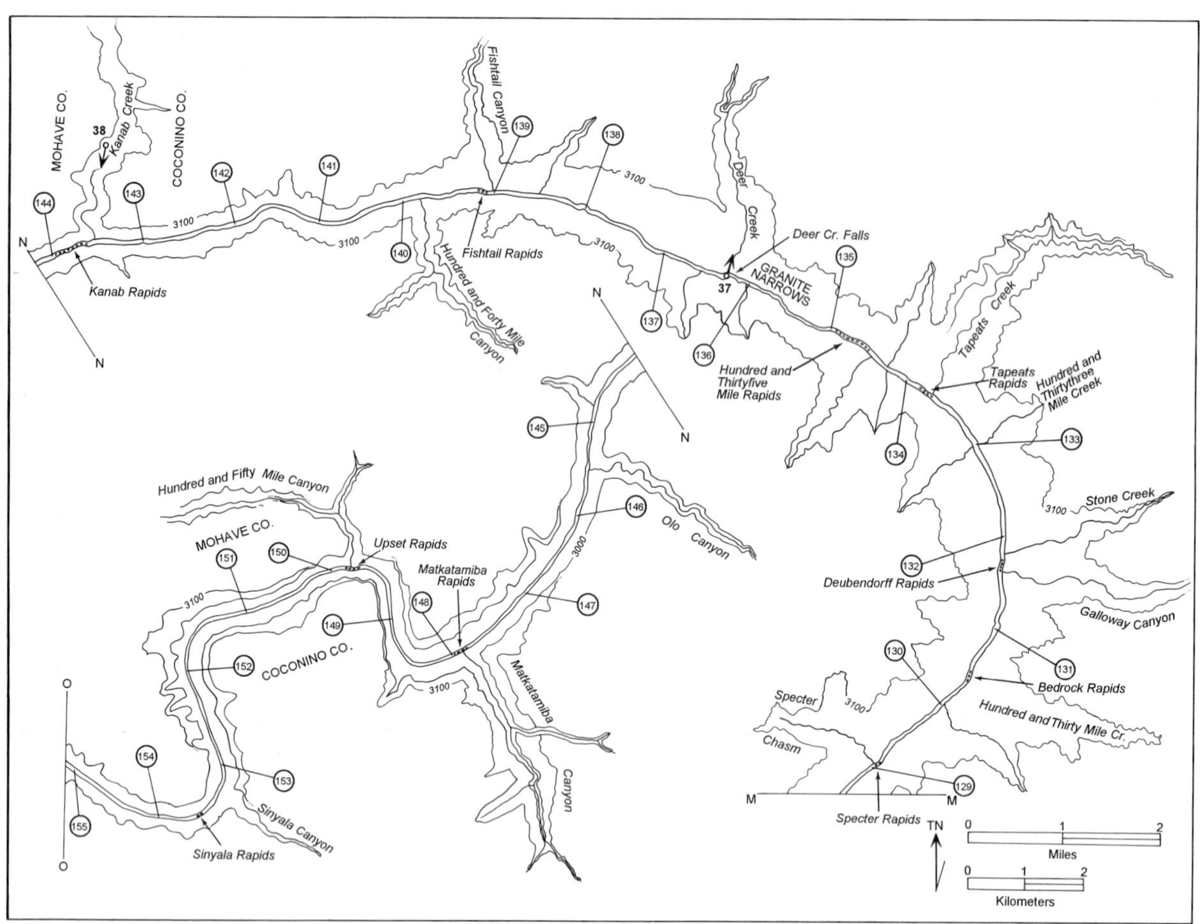

37. Deer Creek Falls, mile 136.3, September 19, 1991, 10:00 A.M.

38. Kanab Creek, 1.5 miles up Kanab Creek from mile 143.5, September 19, 1991, 2:00 P.M.

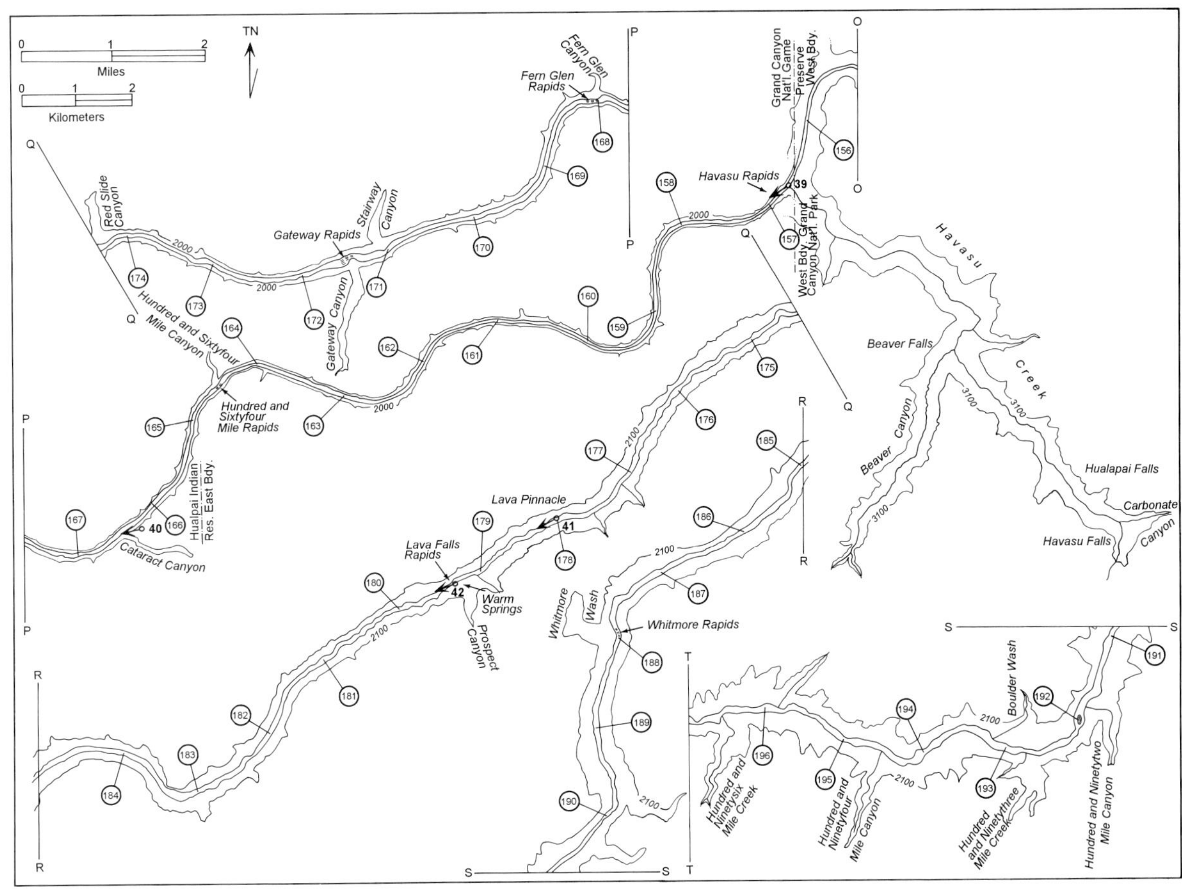

39. Mouth of Havasu Creek, mile 156.7, September 20, 1991, 10:30 A.M.

40. Above National Canyon, mile 166.5, September 20, 1991, 3:00 P.M.

41. Vulcan's Anvil, mile 177.9, September 21, 1991, 10:00 A.M.

42. Lava Falls, mile 179.5, September 21, 1991, 11:00 A.M.

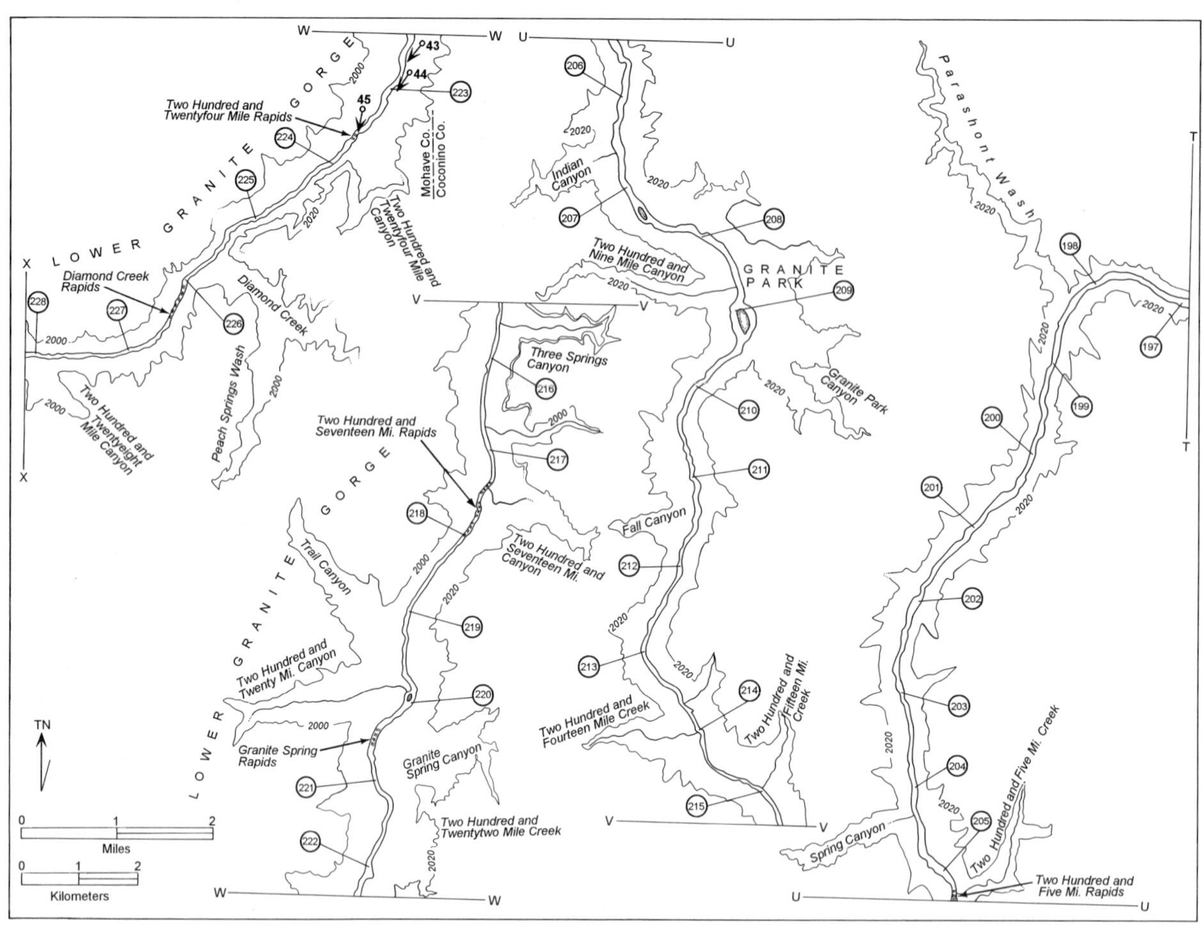

43. Above Diamond Peak, mile 222.5, September 22, 1991, 11:00 A.M.

44. Diamond Peak, mile 222.8, September 22, 1991, 1:00 P.M.

45. 224-Mile Rapid, mile 223.3, September 22, 1991, 3:00 P.M.

Maps indicating photograph locations are based upon the published USGS 1923 Birdseye survey maps, redrafted by Susan Kenzle, P-III Associates, Inc.

BIBLIOGRAPHY

Baars, Donald L. "Major John Wesley Powell: Colorado River Pioneer." In *Geology and Natural History of the Grand Canyon Region,* edited by Donald L. Baars, pp. 10–18. Fifth Field Conference, Powell Centennial River Expedition, Four Corners Geological Society, 1969.

_____. *The Colorado Plateau: A Geologic History.* Albuquerque: University of New Mexico Press, 1983.

Belknap, Buzz, and Loie B. Evans. *Grand Canyon River Guide.* Evergreen, Colorado: Westwater Books, 1990.

Beus, Stanley S., and Michael Morales, eds. *Grand Canyon Geology.* New York/Oxford: Oxford University Press, 1990.

Birdseye, Claude H. *Topographic Instructions of the United States Geological Survey.* USGS Bulletin 788. Washington, D.C.: U.S. Government Printing Office, 1928.

Birdseye, Claude H., and R. W. Burchard. *Plan and Profile of Colorado River from Lees Ferry, Arizona, to Black Canyon, Arizona-Nevada, and Virgin River, Nevada.* Department of Interior/U.S. Geological Survey. Washington, D.C.: U.S. Government Printing Office, 1924.

Birdseye, Claude H., and Raymond C. Moore. "A Boat Voyage through the Grand Canyon of the Colorado." *Geographical Review* 14 (April 1924):177–196.

Blake, H. Elwyn. "Diary Kept by Elwyn Blake While on a Boat Trip through Grand Canyon." 1923. Manuscript in possession of Richard Westwood.

Calvin, William H. *The River that Flows Uphill: A Journey from the Big Bang to the Big Brain.* San Francisco: Sierra Club Books, 1986.

Carothers, Steven W., and Bryan T. Brown. *The Colorado River through the Grand Canyon: Natural History and Human Change.* Tucson: University of Arizona Press, 1991.

Clover, E. U., and Lois Jotter. "Floristic Studies in the Canyon of the Colorado and Tributaries." *American Midland Naturalist* 32:591–642.

Cooley, John R. *The Great Unknown: The Journals of the Historic First Expedition Down the Colorado River.* Flagstaff, Arizona: Northland Publishing, 1988.

Crumbo, K. *A River Runner's Guide to the History of the Grand Canyon.* Boulder, Colorado: Johnson Books, 1981.

Darrah, William C. *Powell of the Colorado.* Princeton, New Jersey: Princeton University Press, 1951.

_____, ed. "Biographical Sketches and Original Documents of the First Powell Expedition of 1869." *Utah Historical Quarterly* 15 (1947):1–148.

Darton, N. H., and others. *Guidebook of the Western United States: Part C. The Santa Fe Route with a Side Trip to the Grand Canyon of the Colorado.* USGS Bulletin 613. Washington, D.C.: U.S. Government Printing Office, 1915.

Dellenbaugh, Frederick S. *The*

Romance of the Colorado River. New York: G. P. Putnam, 1904.

———. *A Canyon Voyage: The Narrative of the Second Powell Expedition.* New Haven, Connecticut: Yale University Press, 1926.

Dutton, C. E. "The Physical Geology of the Grand Canyon District." In *Second Annual Report of the U.S. Geological Survey,* pp. 47–166. Washington, D.C.: U.S. Government Printing Office, 1881.

———. *Tertiary History of the Grand Canyon District, with Atlas.* Washington, D.C.: U.S. Government Printing Office, 1882.

Fradkin, Philip L. *A River No More: The Colorado River and the West.* New York: Knopf, 1981.

Freeman, Lewis R. "Surveying the Grand Canyon of the Colorado." *National Geographic* 45 (May 1924):472–532.

Ghiglieri, Michael. *Canyon.* Tucson: University of Arizona Press, 1992.

Gilbert, G. K. "Report on the Geology of Portions of New Mexico and Arizona." In *U.S. Geographical and Geological Surveys West of the 100th Meridian Report.* Washington, D.C.: U.S. Government Printing Office, 1875.

———. *John Wesley Powell: A Memorial to an American Explorer and Scholar.* Chicago: Open Court Press, 1903.

Hamblin, W. Kenneth. "Late Cenozoic Lava Dams in the Western Grand Canyon." In *Grand Canyon Geology*, edited by S. S. Beus and M. Morales, 385–434. New York: Oxford University Press, 1990.

Hamblin, W. Kenneth, and Joseph R. Murphy. *Grand Canyon Perspectives: A Guide to the Canyon Scenery by Means of Interpretive Panoramas.* Provo, Utah: H & M Distributors, 1969.

Hillers, John K. *Photographed All the Best Scenery: Jack Hillers' Diary of the Powell Expeditions, 1871–75.* Edited by Don Fowler. Salt Lake City: University of Utah Press, 1972.

Ives, Lt. J. C., and J. S. Newberry. *Report upon the Colorado River of the West, Explored in 1857 and 1858.* Washington, D.C.: U.S. Government Printing Office, 1861.

Klett, Mark, and others. *Second View: The Rephotographic Survey Project.* Albuquerque: University of New Mexico Press, 1984.

Kolb, Ellsworth L. *Through the Grand Canyon from Wyoming to Mexico.* New York: Macmillan, 1914.

Kolb, Ellsworth L., and Emery C. Kolb. "Experiences in the Grand Canyon." *National Geographic* 26 (August 1914):99–184.

LaRue, Eugene C. "Water Power and Flood Control of Colorado River below Green River, Utah." In *Water-Supply Paper 556,* pp. 1–123. Department of the Interior/U.S. Geological Survey, Washington, D.C.: U.S. Government Printing Office, 1925.

Lavender, David. *Colorado River Country.* New York: E. P. Dutton, 1982.

———. *River Runners of the Grand Canyon.* Tucson: University of Arizona Press, 1985.

Leydet, F. *Time and the River Flowing: Grand Canyon.* San Francisco: Sierra Club, 1964.

Lucchitta, Ivo. "History of the Grand Canyon and of the Colorado River in Arizona." In *Grand Canyon Geology*, edited by S. S. Beus and M. Morales, 311–332. New York: Oxford University Press, 1990.

Malde, H. E. "Geologic Bench Marks by Terrestrial Photography." USGS *Journal of Research* 1 (1973):193–206.

———. "Panoramic Photographs." *American Scientist* 71 (March–April 1983):132–140.

Moore, Raymond C. "Stratigraphy of a Part of Southern Utah." *Bulletin of the American Association of Petroleum Geologists* 6 (1922):199–227.

———. Field Notebook, 1923. R. C. Moore Collection, University of Kansas Archives, Spencer Research Library, Lawrence, Kansas.

———. "Geologic Report on the Inner Gorge of the Grand Canyon of Colorado River." In *Water-Supply Paper 556,* pp. 125–171 and maps. Department of the Interior/U.S. Geological Survey. Washington, D.C.: U.S. Government Printing Office, 1925.

Moore, Raymond C., C. R. Longwell, H. D. Miser, K. Bryan, and S. Paige. "Rock Formations in the Colorado Plateau of Southeastern Utah and Northern Arizona." In *Shorter Contributions to General Geology 1923–24,* pp. 1–25. U.S. Geological Survey Professional Paper 132. Washington, D.C.: U.S. Government Printing Office, 1925.

Powell, John Wesley. *Exploration of the Colorado River of the West and*

Its Tributaries. Washington, D.C.: U.S. Government Printing Office, 1875. Reprinted Garden City, New York: Doubleday Anchor, 1961.

———. *Report on the Lands of the Arid Region of the United States, with a More Detailed Account of the Lands of Utah.* Washington, D. C.: U.S. Government Printing Office, 1878.

Pyne, Stephen J. *Dutton's Point: An Intellectual History of the Grand Canyon.* Monograph 5. Grand Canyon: Grand Canyon Natural History Association, 1982.

Rogers, Garry F., Harold E. Malde, and Raymond M. Turner. *Bibliography of Repeat Photography for Evaluating Landscape Change.* Salt Lake City: University of Utah Press, 1984.

Stanton, Robert Brewster. *Down the Colorado.* Edited by D. W. Smith. Norman: University of Oklahoma Press, 1968.

Stegner, Wallace. *Beyond the Hundredth Meridian: John Wesley Powell and the Second Opening of the West.* Boston: Houghton Mifflin, 1954.

Stephens, Hal G., and Eugene M. Shoemaker. *In the Footsteps of John Wesley Powell: An Album of Comparative Photographs of the Green and Colorado Rivers, 1871–72.* Boulder, Colorado: Johnson Books, 1987.

Stevens, L. *The Colorado River in Grand Canyon: A Guide.* 3rd ed. Flagstaff, Arizona: Red Lake Books, 1990.

Suran, William C. *The Kolb Brothers of the Grand Canyon.* Grand Canyon, Arizona: Grand Canyon Natural History Association, 1991.

Turner, Raymond M., and Martin M. Karpiscak. *Recent Vegetation Changes along the Colorado River between Glen Canyon Dam and Lake Mead, Arizona.* USGS Professional Paper 1132. Washington, D.C.: U.S. Government Printing Office, 1980.

Webb, R. H., S. S. Smith, and V. A. S. McCord. *Historical Channel Change of Kanab Creek.* Grand Canyon, Arizona: Grand Canyon Natural History Association, 1992.

Westwood, Richard E. *Rough-Water Man: Elwyn Blake's Colorado River Expeditions.* Reno: University of Nevada Press, 1992.

Wheeler, Lt. G. M. *Report upon United States Geographical Surveys West of the One Hundredth Meridian: Geographical Report.* Washington, D.C.: U.S. Government Printing Office, 1889.

Worster, Donald. *Rivers of Empire: Water, Aridity, and the Growth of the American West.* New York: Pantheon, 1985.

SOURCES OF THE HISTORICAL PHOTOGRAPHS

1 USGS

2 KU ARCHIVE
3 KU ARCHIVE
4 KU ARCHIVE
7 KU ARCHIVE
12 KU ARCHIVE
16 KU ARCHIVE
19 KU ARCHIVE
20 KU ARCHIVE
21 KU ARCHIVE
23 KU ARCHIVE
24 KU ARCHIVE

Emery Kolb Collection at Northern Arizona University Photo Numbers for *The Canyon Revisited*

5 NAU 568-10934
6 NAU 568-5142
8 NAU 568-5098
9 NAU 568-5033
10 NAU 568-5162
11 NAU 568-3210
13 NAU 568-5067
14 NAU 568-5097
15 NAU 568-5196
17 NAU 568-3284
18 NAU 568-3213
22 NAU 568-5100
25 NAU 568-10947
26 NAU 568-5086
27 NAU 568-3248
28 NAU 568-4993
29 NAU 568-5195
30 NAU 568-6075
31 NAU 568-5194
32 NAU 568-5003
33 NAU 568-2981
34 NAU 568-3224
35 NAU 568-5025
36 NAU 568-4986
37 NAU 568-5043
38 NAU 568-5062
39 NAU 568-3234
40 NAU 568-3243
41 NAU 568-3328
42 NAU 568-5185
43 NAU 568-3268
44 NAU 568-6071
45 NAU 568-5014